Conversion of
Large Scale Wastes into
Value-added Products

Conversion of Large Scale Wastes into Value-added Products

Edited by
Justin S. J. Hargreaves   Ian D. Pulford
Malini Balakrishnan   Vidya S. Batra

CRC Press
Taylor & Francis Group
Boca Raton London New York

CRC Press is an imprint of the
Taylor & Francis Group, an **informa** business

CRC Press
Taylor & Francis Group
6000 Broken Sound Parkway NW, Suite 300
Boca Raton, FL 33487-2742

First issued in paperback 2016

Version Date: 20131031

ISBN 13: 978-1-138-19880-7 (pbk)
ISBN 13: 978-1-4665-1261-0 (hbk)

Visit the Taylor & Francis Web site at
http://www.taylorandfrancis.com

and the CRC Press Web site at
http://www.crcpress.com

Contents

Preface

The application of waste as a resource is the subject of increasing attention worldwide. This is driven by a combination of factors, which include a growing appreciation of the need to develop more sustainable processes and to utilize Earth's finite resources more efficiently as well as the reduction of waste storage requirements, which can be significant, and even potentially hazardous, in the case of some large-scale processes. In this book, the utilization, and potential utilization, of a number of selected large-scale wastes is described, providing a snapshot of an area that is continually evolving. It is important to note that this area comprises a number of well-established and widely practiced processes in addition to the drive toward the development of strategies for the application of wastes in new ways or the use of wastes not previously applied. The nature of certain types of waste and routes for their application is also subject to a degree of geographical variation. Chapter 1 is a general introduction to the area of large-scale waste utilization. In Chapter 2, the nature of the various types of waste generated from different large-scale metal processing operations is described along with their application or potential application. Chapter 3 outlines waste generated by coal combustion, a major source of power generation, in which enormous quantities of waste are generated and some of which is applied in the manufacture of construction materials. The subject of waste electrical and electronic equipment, an area increasingly recognized as important particularly for the recycling of finite resources, is discussed in Chapter 4. The application and potential application of food waste, which is a significant but very diverse waste stream, is described in Chapter 5. Chapter 6 provides a general conclusion to the wide-ranging area of waste utilization. Future developments in the area of waste utilization can be anticipated.

In bringing this book to fruition, we would like to extend our gratitude to Professor Jack Groppo (University of Kentucky, USA) and Dr. Nicola Dunn (National Farmers' Union, UK) for kindly contributing Chapters 3 and 5, respectively. We would also like to thank the staff at CRC Press, Taylor & Francis Group, in particular Hilary Rowe and David Fausel, for all their generous assistance.

M. Balakrishnan and V.S. Batra
TERI, New Delhi, India

J.S.J Hargreaves and I.D Pulford
University of Glasgow, UK

Contributors

M. Balakrishnan
The Energy and Resources Institute
  (TERI)
New Delhi, India

V.S. Batra
The Energy and Resources Institute
  (TERI)
New Delhi, India

N. Dunn
NFU, Agriculture House
Warwickshire, UK

J.G. Groppo
Environmental & Coal Technologies
Center for Applied Energy Research
University of Kentucky
Lexington, KY

J.S.J. Hargreaves
School of Chemistry
University of Glasgow
Glasgow, UK

I.D. Pulford
School of Chemistry
University of Glasgow
Glasgow, UK

Abbreviations

ABP	Animal By-Product
ABS	Acrylonitrile-Butadiene-Styrene
ACAA	American Coal Ash Association
ACI	Activated Carbon Injection
AD	Anaerobic Digestion
ADAA	Ash Development Association of Australia
AFBC	Atmospheric Fluidized Bed Combustion
AMD	Acid Mine Drainage
APP	Asia Pacific Partnership
ASA	Acrylester Styrene Acrylonitrile
ASTM	American Society for Testing and Materials
BF	Blast Furnace
BOF	Basic Oxygen Furnace
BRIC	Brazil, Russia, China, India
CCB	Coal Combustion By-Products
CDF	Controlled Density Fill
CFBC	Circulating Fluidized Bed Combustion
CFB-FGD	Circulating Fluidized Bed-Flue Gas Desulfurization
CFC	Chlorofluorocarbon
CLSM	Controlled Low-Strength Material
CMU	Concrete Masonry Units
CRT	Cathode Ray Tube
CSA	Calcium Sulfoaluminate Cements
DRMMC	Discontinuously Reinforced Metal-Matrix Composites
EAF	Electric Arc Furnace
EEE	Electrical and Electronic Equipment
EPR	Extended Producer Responsibility
ESP	Electrostatic Precipitators
EU	European Union
FAO	Food and Agriculture Organization
FBC	Fluidized Bed Combustion
FGD	Flue Gas Desulfurization
GCC	Gulf Cooperation Council
GHG	Greenhouse Gas
HCFC	Hydrochlorofluorocarbon
HIPS	High-Impact Polystyrene
LCA	Life Cycle Assessment
LCD	Liquid Crystal Display
LCLL	Low Caustic Leaching and Liming
LOI	Loss on Ignition
MBT	Mechanical Biological Treatment
NMP	Non-Metallic Products

NMR	Non-Metallic Residue
OECD	Organization for Economic Cooperation and Development
OPC	Ordinary Portland Cement
PA	Polyamide
PBB	Polybrominated Biphenyl
PBDE	Polybrominated Diphenyl Ether
PC	Polycarbonate
PC	Pulverized Coal
PCB	Polychlorinated Biphenyl
PE	Polyethylene
PFBC	Pressurized Fluidized Bed Combustion
PGE	Platinum Group Element
PP	Polypropylene
PPE	Polyphenylene-Ether
PPO	Polyphenylene Oxide
PS	Polystyrene
PSM	Pozzolanic-Stabilized Mixture
PU	Polyurethene
PVC	Polyvinyl Chloride
RCRA	Resource Conservation and Recovery Act
REE	Rare Earth Element
RoHS	Restriction of Use of Certain Hazardous Substances
SAN	Styrene-Acrylonitrile
SCR	Selective Catalytic Reduction
SDA	Spray Drier Absorber
SNCR	Selective Non-Catalytic Reduction
SPL	Spent Pot Lining
USEPA	United States Environmental Protection Agency
WEEE	Waste Electrical and Electronic Equipment
WFD	Waste Framework Directive
WRAP	Waste and Resources Action Programme

1 Introduction

M. Balakrishnan, V.S. Batra,
J.S.J. Hargreaves, and I.D. Pulford

CONTENTS

Recently, concern about the fate of waste products produced by a whole range of industrial processes has combined with a growing realization that they may have potential uses and, therefore, value (Balakrishnan et al., 2011; Gupta et al., 2009). This has led to a large number of studies aimed at utilizing such wastes, which is the focus of this book. The following chapters discuss various selected classes of large-scale waste and their current applications and potential future applications. Chapter 2 details different wastes derived from metal processing. In Chapter 3, combustion products are discussed; here, the focus is on coal combustion products given the widespread use of coal. Chapter 4 details waste electrical and electronic equipment (WEEE) and Chapter 5 discusses food waste. Before discussing each of these classes, it is important to make a number of general considerations.

1.1 DRIVERS FOR WASTE RECOVERY AND REUSE

A number of factors combine to act as drivers for waste recovery and reuse:

1. Problems of disposal of large amounts of waste
2. Environmental issues
3. Availability of resources
4. Economics
5. Legislation

The focus widens going down this list, from aspects that affect mainly the producers of the waste to ones affecting the whole community. The environmental and economic issues drove the political requirements that most developed, and increasingly less-developed, countries have now translated into legislation.

1.1.1 PROBLEMS OF DISPOSAL OF LARGE AMOUNTS OF WASTE

Some industries produce very large amounts of waste. For example, global fly ash production from coal combustion (Chapter 3) is estimated to be of the order of 500 million tonnes/year, while slag from iron and steel production and red mud from production of aluminium (Chapter 2) both generate about 100 million tonnes/year. (Rounded figures are used here; see individual chapters for more detail). As these are continuous processes, the wastes must be removed from the systems in order for them to function efficiently. Traditionally, such wastes have been dumped in heaps or lagoons, which can become extensive as more wastes are produced. The disposal sites also have to be close to the source of the waste (blast furnace, power station, etc.) in order to reduce transportation costs. As a result, large areas adjacent to such facilities can be given over to disposal and storage of waste.

Waste from various food producing processes can also be significant (Chapter 5). For example, approximately 60 to 70 million tonnes of both rice husk and orange peel are produced annually. Again, such waste is produced locally where foodstuffs are processed. For example, it is estimated that 2.5 to 4.0 million tonnes of olive oil processing waste are produced annually in the Andalusia region of Spain alone.

Over recent years, WEEE (Chapter 4) has become significant, with an estimated annual production of 20 to 25 million tonnes. Much of the material that goes into this waste stream is bulky (refrigerators, television sets, etc.), which causes particular storage problems.

In addition to the availability of land on which to store wastes, the main issues that also need to be considered are safety and toxicity. Storage of large volumes of waste presents the problem of physical containment. In some cases, the waste may contain toxic components, whose dispersal into the wider environment could be detrimental to human health and the ecology of the local area. There have been a number of high-profile incidents over recent years where failure of the waste containment system has led to fatalities, poisoning of humans, and contamination of land and water. Selected recent incidents taken from a list of major tailings dam failures spanning the previous 50+ years published on the Web (http://www.wise-uranium.org/mdaf.html)

are presented in Table 1.1. From the table, the damage wrought to the local environment, and in some cases even human life, is apparent. The scale of the wastes discharged in the incidents is noteworthy and provides an environmental context for the drive to utilize them in a more profitable manner.

1.1.2 Environmental Issues

As well as the contamination issues outlined above, reuse of wastes has implications for resource depletion and energy use. In some cases, the reuse of waste materials could produce savings in that new resources would not be exploited, and the energy used to mine and refine them could be saved. Long-term supply of certain resources is becoming increasingly uncertain. In 2010, the EU identified 14 raw materials whose supply was considered becoming critical, either due to the known reserves of these resources being used up or due to uncertainty of supply because of political and social conditions in the main areas of supply (UNEP, 2009). It has been estimated that recovery of components in WEEE could meet future requirements for many of the precious metals used in the manufacture of electrical and electronic products. Furthermore, aluminium produced by recycling WEEE would require less than 10% of the energy required to produce the equivalent amount by smelting bauxite. There would also be lower emissions of CO_2 and SO_2, and no associated production of red mud (see Chapter 4).

1.1.3 Resource Recovery

Resource recovery has become a major issue as the demand for certain materials has grown over recent years. The supply of some metals in particular has been identified as critical. In a UNEP report in 2009, a timescale was established to prioritize which metals would become critical due to a combination of growth in demand, risks to supply, and restrictions on recycling (UNEP, 2009). It was estimated that supply of tellurium, indium, and gallium would be critical by 2014; the rare earth elements, platinum group elements, lithium, and tantalum by 2020; and germanium and cobalt by 2050. This report recommended three areas that could be developed in order to alleviate the supply of these critical metals:

- the enlargement of recycling capacities
- the development and realization of new recycling technologies
- the accelerated improvement of international recycling infrastructures

In addition to the recovery of raw materials, there is also considerable interest in using wastes as a source of other chemicals, and so precluding the need for their manufacture, with consequent savings in costs, materials, and energy. Food wastes have been suggested as a source of platform chemicals that can be used in the manufacture of other materials, such as pharmaceuticals, food additives, etc. Specific wastes, such as citrus fruit waste, can be a source of high-value products such as essential oils and enzymes (see Chapter 5). The plastic component of WEEE can be a source of monomers for polymer production (see Chapter 4).

TABLE 1.1

Selected Recent Accidents Related to the Storage of Large-Scale Wastes as Published on the Web

Date	Location	Incident	Contaminant	Effect
November 2012	Sotkamo, Kainuu Province, Finland	Leak from gypsum pond	Hundreds of thousands of cubic meters of contaminated waste water	Nickel and zinc concentrations in nearby Snow River exceeded values that are harmful to organisms at least tenfold
October 2010	Kolontár, Hungary	Tailings dam failure	700,000 m³ of caustic red mud	Several towns flooded, 10 people killed, approximately 120 people injured, 8 km² flooded
June 2010	Huancavelica, Peru	Tailings dam failure	21,420 m³ of tailings	Contamination of Rio Escalera and Rio Opamayo 100 km downstream
May 2009	Huayuan County, Xiangxi Automonous Prefecture, Hunan Province, China	Tailings dam failure (capacity 50,000 m³)		The landslide set off by the tailings dam failure destroyed a home, killing 3 people and injuring 4 people
December 2008	Harriman, Tennessee	Retention wall failure	4,100,000 m³ of coal ash slurry	1.6 km² covered as deep as 1.83 m, 12 homes damaged
November 2004	Pinchi Lake, British Columbia, Canada	Tailings dam collapse	6000–8000 m³ of rock, dirt, and waste water	Tailings spilled into Pinchi Lake

May 2004	Partizansk, Primorski Krai, Russia	Breakage of ring dike	160,000 m^3 of coal ash	The ash flowed through a drainage canal into a tributary to the Partizanskaya River, which empties into Nahodka Bay in Primorski Krai
March 2004	Malvési, Aude, France	Dam failure after heavy rain in preceding year	30,000 m^3 of liquid and slurries	Elevated nitrate concentrations up to 170 mg/L in the Tauran Canal for several weeks
October 2003	Quinta Region, Chile	Tailings dam failure	50,000 tonnes of tailings	Tailings flowed 20 km downstream the Rio La Ligua
January 2000	Baia Mare, Romania	Tailings dam failure	100,000 m3 of cyanide contaminated liquid	Contamination of the Somes River, a tributary of the Tisza River, killing fish and poisoning the drinking water of >2 million people in Hungary

Source: http://www.wise-uranium.org/mdaf.html.

1.1.4   ECONOMICS

While the issues outlined previously are important in their own right, they also affect the economics of production. Some of the economic benefits are linked to lower energy requirements as in the case of aluminium recycling. Compared to steel making from iron ore, its production from scrap leads to economic benefits arising from 74% reduction in energy requirement and 40% reduction in water requirement (Arbatov). In the cement sector as well, economic benefits are obtained with the use of waste biomass based fuels instead of coal, and use of waste (fly ash, slag) as partial substitute for clinker thus saving on energy required for clinker production (Lafarge, 2008). In addition to economic benefits from reduction in resource requirements, costs related to handling and managing the waste (disposal tax, tax on primary construction material, etc.) are also saved (Harrison et al., 2002). The waste is instead converted to a by-product or raw material that generates additional revenue. Several utilization options also address the negative properties of the waste (binding of heavy metals, neutralization of high alkalinity) and thus the potential health costs due to exposure and clean-up costs in case of accidents are saved as well.

1.1.5   LEGISLATION

The growing realization of the economic and environmental benefits of recycling and reusing wastes has led to a wealth of legislation across the world dealing with waste prevention, minimization, recovery, and reuse. The Basel Convention homepage provides a summary of waste legislation by country as of October 2011 (http://www.basel.int/Countries/Countryfactsheets/tabid/1293/Default.aspx).

The European Union (EU) has passed a series of laws relating to waste designed to provide environmental protection. The overarching legislation is the Waste Framework Directive (WFD), with specific legislation dealing with hazardous waste, waste from consumer goods, waste from specific activities, and radioactive waste and substances. Table 1.2 shows the range of legislation, which can be accessed at http://europa.eu/legislation_summaries/environment/waste_management/index_en.htm

Within the EU, individual countries translate the policy into their own legislation. For example, the UK has The Controlled Waste (England & Wales) Regulations 2012 (http://www.legislation.gov.uk/uksi/2012/811/contents/made) and The Waste (Scotland) Regulations 2012 (http://www.legislation.gov.uk/sdsi/2012/9780111016657/contents).

In the United States, the federal Resource Conservation and Recovery Act (RCRA) (http://www.epa.gov/osw/laws-regs/) is the primary law governing the disposal of solid and hazardous waste. The United States Environmental Protection Agency (USEPA) delegates the primary responsibility of implementing the RCRA hazardous waste program to individual states so that minimum standards are applied consistently across the country, but individual states have flexibility in implementing the rules. State RCRA regulations must be at least as stringent as the federal requirements, but states have the power to adopt requirements that are more stringent.

TABLE 1.2
EU Legislation Dealing with Wastes

General Framework

Directive on waste
Waste management statistics
Landfill of waste
Waste incineration
Shipments of waste
Strategy on the prevention and recycling of waste (Archives)
The management of bio-waste in the European Union (Archives)

Hazardous Waste

Basel Convention
Controlled management of hazardous waste (until the end of 2010) (Archives)

Waste From Consumer Goods

Packaging and packaging waste
Disposal of polychlorinatedbiphenyls (PCBs) and polychlorinatedterphenyls (PCTs)
Disposal of spent batteries and accumulators
End-of-life vehicles
The reusing, recycling, and recovering of motor vehicles
Waste electrical and electronic equipment (WEEE)
Substances subject to restrictions for use in electrical and electronic equipment

Waste From Specific Activities

Industrial emissions
Integrated pollution prevention and control (until 2013)
Management of waste from extractive industries
A strategy for better ship dismantling practices
Removal and disposal of disused offshore oil and gas installations
Use of sewage sludge in agriculture
Port facilities for ship-generated waste and cargo residues
Titanium dioxide
  • Disposal of titanium dioxide industrial waste
  • Surveillance and monitoring of titanium dioxide waste
  • Reduction of pollution caused by waste from the titanium dioxide industry

Radioactive Waste and Substances

Shipments of radioactive waste: supervision and control
Shipments of radioactive substances
Situation in 1999 and prospects for radioactive waste management
Management of spent fuel and radioactive waste

Source: http://europa.eu/legislationsummaries/environment/wastemanagement/indexen.htm.

In Canada, the Canadian Environmental Protection Act of 1999 is the primary legislation, which contains sections dealing specifically with pollution prevention, management of toxic substances, controlling pollution, and managing wastes (http://www.ec.gc.ca/p2/default.asp?lang=En&n=CC0A793B-1). Australia's National Waste Policy (http://www.environment.gov.au/wastepolicy/index.html) is designed to

> avoid the generation of waste and reduce the amount of waste (including hazardous waste) for disposal; manage waste as a resource; ensure that waste treatment, disposal, recovery and re-use is undertaken in a safe, scientific and environmentally sound manner; and contribute to the reduction in greenhouse gas emissions, energy conservation and production, water efficiency and the productivity of the land.

In New Zealand, the focus is on waste minimization with the Waste Minimisation Act of 2008 (http://www.mfe.govt.nz/issues/waste/waste-minimisation.html). In South Africa, the National Waste Management Strategy (http://www.wastepolicy.co.za/) is underpinned by the Waste Act, 2008 (http://www.sawic.org.za/?menu=13).

In India, there are several national regulations pertaining to waste management (http://www.indiawastemanagementportal.org/index.php?option=com_content&view=article&id=76&Itemid=162; Galea, 2010). The main legislation to regulate all forms of waste is the Environment Protection Act of 1986; the corresponding Environmental Protection Rules 1986 provide the details. The latter can be modified without amending the principal act. Various aspects of hazardous wastes covering handling, storage, reprocessing, and reuse are governed by The Hazardous Wastes (Management, Handling and Transboundary Movement) Rules, 2008. Moreover, there are separate rules for plastics and E-waste, namely, The Plastics (Manufacture, Usage and Waste Management) Rules, 2009 and The E-Waste (Management and Handling Rules), 2010, respectively.

Comprehensive resource utilization was included as a main aim of environmental protection in the Environmental Protection Law of China promulgated in 1979 (Wang, 1998). Subsequently, the law on Prevention and Control of Environmental Pollution Caused by Solid Waste of PRC was enacted on April 1, 1996. In 2003, the Law for Promotion of Cleaner Production of PRC came into force (Hoornweg, Lam, and Chaudhry, 2005), followed in 2009 by the Circular Economy Promotion Law that emphasizes the 3Rs (reduce, reuse, and recycle) (UNCRD, 2011). As part of the Chinese Government's 12th five-year plan, the utilization of industrial solid waste as a secondary resource is also being encouraged (Song and Jiang, 2012).

1.2 TYPES OF WASTE

Wastes can be broadly categorized into mineral and organic materials. The main mineral wastes are produced by coal combustion (Chapter 3) and metal mining and processing (Chapter 2), with WEEE now also being a significant source (Chapter 4). Organic wastes are mainly from food and food processing (Chapter 5). In both cases, disposal has traditionally been by dumping, either into specifically designed areas or into landfill. Organic wastes have the advantage that, at least theoretically, they can be decomposed, while mineral wastes cannot.

1.3 SECONDARY PROCESSES

1.3.1 CATALYSIS

The application of large-scale wastes as catalytic resources has been the focus of a recent review (Balakrishnan et al., 2011) and a special issue of a catalysis journal (Hargreaves and Batra, 2012). Within this area, wastes have been applied directly as catalysts themselves (e.g., in the direct cracking of methane over red mud pre-catalysts; Balakrishnan et al., 2009) or as resources from which components can be extracted for the further synthesis or application as materials (e.g., the extraction of SiO_2 from rice husk as reviewed by Adam, Appaturi, and Iqbal, 2012). In the latter regard, procedures can be adopted for the enhancement of the concentration and/ or properties of a particular component. For example, as described in Chapter 2, pre-treatment procedures are often applied to red mud, with the aim of enhancing the catalytic properties of the iron component (Sushil and Batra, 2008). Table 1.3 summarizes a number of recent studies in which the application of wastes directly as catalysts or as sources from which components of catalysts have been derived has been reported.

By definition, the application of wastes or waste-derived components as catalytic resources will only form a component of large-scale waste utilization since it is unlikely to provide a continuous high-volume demand. Consideration also has to be given to the performance of the materials as set against those derived from more traditional routes. Sometimes a balance between poorer efficacy, greater resource efficiency, legislation, and lower cost may be necessary.

TABLE 1.3
Selected Wastes Applied As Sources of Catalysts and Related Materials

Red mud	Balakrishan et al., 2009; Sushil et al., 2010; Karimi et al., 2013; De Resende et al., 2013; Liu et al., 2013; Paredes et al., 2004; Saptura et al., 2012; Karimi et al., 2012; Sushil and Batra, 2012
Fly ash	Rode et al., 2012; Rios R., 2012; Saptura et al., 2012; Musyoka et al., 2012; Pande et al., 2012
Bottom ash	Chiang et al., 2012; Park et al., 2012
Slag	Kuwahara and Yamashita, 2013; Kuwahara et al., 2010; Kuwahara et al., 2012
Aluminium dross	Kim et al., 2009; Muarayama, 2006
Rice husk ash	Adam et al., 2012; Rafiee et al., 2012; Bin Shawkataly et al., 2012
Chitosan	Schussler et al., 2012; Zeng et al., 2012; Macquarrie and Hardy, 2005
Egg shells	Gao and Xu, 2012; Khemthong et al., 2012; Wei et al., 2009; Viriya-empikul et al., 2010; Cho and Seo, 2010
Cockle shells	Boey et al., 2012
Biochar	Kastner et al., 2012; Ormsby et al., 2012

1.3.2 ENERGY PRODUCTION

Historically, the term "waste to energy" has referred to the combustion of rubbish. However, these days, there are a number of different technologies that can be applied, as discussed elsewhere (http://www.e-renewables.com/documents/Waste/Waste%20to%20Energy%20-%20The%20Basics.pdf). These include physical technologies in which waste is processed to make it more useful as a fuel, thermal processes that yield heat or fuel, and bacterial processes in which organic wastes are fermented to produce fuel. In physical techniques, pre-processing of primarily municipal waste prior to its application as a fuel is undertaken; this can involve sorting and autoclaving in the presence of high-pressure steam, which can soften some components such as plastics and fibrous material. While this step does add to the processing costs, it results in more efficient and cleaner combustion processes. Thermal processes involve the direct combustion of waste to produce heat as a source of energy, although increasing attention is being directed toward pyrolysis and thermal gasification (two related approaches, the former being carried out anaerobically and the latter in the presence of low levels of oxygen). Pyrolysis can produce gases, liquids, and solid products. The gaseous components can be combusted as a source of energy, and liquid products such as bio-oil can be applied as fuels, provided appropriate pre-treatment is undertaken (e.g., the acidity and other properties of bio-oil produced by fast pyrolysis can be improved by a pre-treatment procedure applying red mud [Karimi, Kalbasi, and Hajrasuliha, 2012]). Biochar, an example of a solid product arising from pyrolysis, can find uses in other applications such as water treatment. Plasma arc gasification can be used to convert waste into synthetic gas and slag. The synthetic gas can then be applied to the production of more traditionally applied hydrocarbon-based fuels. Biological processes can occur naturally; for example, the production of methane (which can be used as fuel) from landfill sites or in intentionally built plants where the digestion of food-processing or agricultural waste to yield product fuel gases can be undertaken in anaerobic digesters inoculated with various types of bacteria. Fermentation can also be undertaken using yeast in which ethanol, a potential fuel, is a target product.

Other potentially useful products can be obtained from the application of wastes as sources of energy or fuel. In addition to biochar mentioned previously, rice husk ash (Adam et al., 2012) and bottom ash generated from the incineration of municipal waste (Chiang et al., 2012) have found application as sources of material for the preparation of useful materials such as in catalysts and water treatment agents.

1.3.3 AGRICULTURAL

Wastes have commonly been added to soil for three main reasons:

* Supply of nutrients
* Improvement of physical characteristics
* Adjustment of pH

Traditionally, manures and slurries have been recycled to soil (Chambers et al., 2001). More recently, sewage sludge, the waste product of sewage treatment plants, has also been used, although there are some concerns regarding the content of heavy metals or other toxins in some sludges (Smith, 1996). The main aim of adding such wastes to soil is to utilize their nutrient content, especially nitrogen, phosphorus, and potassium. The amount of nutrient varies depending on the source and type of waste, but typical contents are 1 to 2% N, 0.5 to 1% P, and 0.5 to 1.0% K. The potassium tends to be immediately available for plant use, but the nitrogen and phosphorus are released more slowly by microbial action. The organic nature of these wastes is beneficial for improvement of the physical structure of soil due to interactions with the mineral component.

Where large quantities of specific wastes are produced, there is considerable interest in their potential reuse for soil improvement; for example, olive oil waste (Arvanitoyannis and Kassaveti, 2007), orange peel (Tuttobene et al., 2009), coffee wastes (Kasongo et al., 2011; Murthy and Naidu, 2012), fly ash (Jala and Goyal, 2006), and flue gas desulfurization product (Balingar et al., 2011). While organic wastes of this type may add nutrients, very often their main benefit is the improvement of soil structure. The inorganic wastes tend to be used to supply the major cations, such as Ca^{2+} and Mg^{2+}, some trace elements, and for pH adjustment.

1.3.4 ENVIRONMENTAL CLEAN-UP

The focus of using wastes for environmental clean-up has been primarily for treating contaminated water (Ahmaruzzaman, 2011; Babel and Kurniawan, 2003; Gupta et al., 2009) and soil (Gadepalle et al., 2007; Kumpiene, Lagerkvist, and Mauruce, 2008; van Herwijnen et al., 2007). Most attention has been paid to the removal from water or stabilization in soil of heavy metals, but there is also some interest in removal of organic contaminants, such as dyes (Ali et al. 2012; Gupta and Suhas, 2009). In many cases the low cost of the waste has been stressed (Ali, Asim, and Khan, 2012; Ahmaruzzaman, 2011; Babel and Kurniawan, 2003; Gupta et al., 2009; Gupta and Suhas, 2009). The effectiveness of the waste materials in removing contaminants tends to be based on one of, or a combination of, three processes: ion exchange, adsorption, and chelation. A large number of wastes have been used for water or soil treatment (Table 1.4).

Very often, the waste material is used relatively unaltered, perhaps only being crushed to increase surface area. In some cases, wastes are treated to improve their sorptive properties. Treatments include acid (e.g., Zhu, Fan, and Zhang, 2008), alkali (e.g., Šćiban, Klašnja, and Škrbić, 2006, 2008), formaldehyde (e.g., Šćiban et al., 2006, 2008), and pelletization (e.g., Kamari, Pulford, and Hargreaves, 2011). There is also much interest in preparing carbons from wastes, which tend to be mesoporous and have higher sorptive properties. This may be done by pyrolizing a waste material, a process recently reviewed by Ali et al. (2012), or by transformation of a waste material used for another purpose; for example, the carbonaceous material produced when red mud has been used to catalyze the cracking of hydrocarbons (Pulford et al., 2012).

TABLE 1.4

Waste Materials Used For Water and Soil Treatment

Type of Waste	Ref.
Bagasse fly ash	Gupta et al., 2003; Shah et al., 2012; Shah et al., 2013
Blast furnace slag	Oguz, 2004; Park et al., 2008; Bowden et al., 2009; Xue et al., 2009; Liu et al., 2010; Ahmaruzzaman, 2011; Zhou and Haynes, 2011
Blast furnace sludge	Lopez et al., 1995; Oritz et al., 2001; Kim et al., 2006; Bhattacharya et al., 2008; Ahmaruzzaman, 2011; Karimian et al., 2012
Bone char	Wilson et al., 2003; Danny et al., 2004; Pan et al., 2009
Brewery waste biomass	Chen and Wang, 2008; Zhang et al., 2010; Wu et al., 2012
Cocoa shells	Meunier et al., 2003; Theivarasu and Mylasamy, 2012
Coconut wastes/coir	Amuda et al., 2007; Conrad and Hansen, 2007; Namasivayam and Sureshkumar, 2008; Nwachukwu and Pulford, 2008
Coffee waste	Oliveira et al., 2008a,b; Kyzas et al., 2012
Fly ash	Ahmaruzzaman, 2011; Zhou and Haynes, 2011
Nut shell	Bulut and Zeki, 2007; Doğan et al., 2008; Pehlivan et al., 2009; Sivrikaya et al., 2012
Orange peel	Sivaraj et al., 2001; Schiewer and Patel, 2008; Feng et al., 2009
Peanut waste	Brown et al., 2000; Gong et al., 2005; Al-Othman et al., 2012
Red mud	Apak et al., 1998; Pradhan et al., 1999; Gupta et al., 2001; Gupta and Sharma, 2002; Santona et al., 2006; Wang et al., 2008; Liu et al., 2011; Zhou and Haynes, 2011; Pulford et al., 2012
Rice husk	Tarley and Arruda, 2004; Vadivelan and Kumar, 2005; Singh and Singh, 2012
Sawdust and wood wastes	Shukla et al., 2002; Šćiban et al., 2006; Nwachukwu and Pulford, 2008
Shellfish waste	Crini and Badot, 2008; Kamari et al., 2011
Tea waste	Amarasinghe and Williams, 2007; Hameed, 2009

1.4   SECONDARY PRODUCTS

1.4.1   CEMENT, CONCRETE, AND CERAMICS

The application of fly ash as a component in construction materials is well established where it could function as a partial replacement for Portland cement and sand. Red mud has also been investigated for construction materials as detailed in a number of studies (e.g., Tsakiridis, Agatzini-Leonardou, and Oustadakis, 2004) and in Chapter 2 although, as discussed, in some cases attention must be paid to the potential radioactivity of materials used for domestic construction purposes (Gu, Wang, and Liu, 2012). Attention has also been directed toward the inclusion of other wastes in cements and concretes such as calcium carbide residue and bagasse ash (Rattanashotinut, Thairit, and Tangchirapat, 2013).

Large-scale wastes can also be applied to the production of ceramics; for example, red mud can be mixed with clay for this purpose (Pontikes and Angelopoulos, 2009).

1.4.2 CHEMICALS

The plastics components of WEEE are used as a source for certain polymers and monomers (see Chapter 4, Schlummer et al., 2006). Food wastes (Chapter 5) can be a source of specific chemicals or groups of chemicals. For example, Siles López, Li, and Thompson (2010) suggested that waste orange peel could be a source of D-limonene (an essential oil that is widely used in flavorings and the chemical industry) and pectin (a gelling agent used in the food industry). Murthy and Naidu (2012) have suggested that coffee wastes could be a source of citric and gibberellic acids, and bio-ethanol.

1.4.3 COMPOSTS

Composting, the breakdown of biodegradable organic material by microbial action, is commonly used for large-scale wastes such as green waste (garden waste) (Benito et al., 2005; Belyaeva, Haynes, and Sturm, 2012), food waste (municipally collected food waste or food industry wastes) (Hargreaves, Adl, and Warman, 2008), and animal manure (Bernal, Alburquerque, and Moral, 2009) (see Chapter 5). There is also considerable interest in the potential use of other wastes in composts; for example, olive oil waste (Roig, Cayuela, and Sánchez-Monedero, 2006; Cayuela, Sánchez-Monedero, and Roig, 2010); coffee wastes (Murthy and Naidu, 2012); tea wastes (Wells et al., 2012); citrus waste (Gelsomino et al., 2010); and seafood waste (Archer and Baldwin, 2006). Co-composting of wastes is commonly carried out in order to promote microbial degradation (Paredes et al., 2002; Belyaeva et al., 2012).

1.4.4 METAL RECOVERY

Recovery of metals from WEEE is commonly undertaken (see Chapter 4) (Cui and Zhang, 2008). Three main groups are recovered: ferrous metal, aluminium, and precious metals. Recovery of metals from wastes like slag, dross, etc., generated in various metallurgical operations are also established (as with aluminium dross) or under exploration.

1.5 CONCLUSION

The following chapters detail aspects of the application of some large-volume wastes, which is an area of growing importance. In these chapters, the diverse nature of the various waste streams is illustrated along with the various stages of their implementation.

REFERENCES

Adam, F., Appaturi, J. N, and Iqbal, A., 2012. The utilization of rice husk silica as a catalyst: review and recent progress. *Catalysis Today*, **190**, 2–14.

Ahmaruzzaman, M., 2011. Industrial wastes as low-cost potential adsorbents for the treatment of wastewater laden with heavy metals. *Advances in Colloid and Interface Science*, **166**, 36–59.

Ali, I., Asim, M., and Khan, T.A., 2012. Low cost adsorbants for the removal of organic pollutants from wastewater. *Journal of Environmental Management*, **113**, 170–183.

Al-Othman, Z.A., Ali, R., and Nanshad, M., 2012. Hexavalent chromium removal from aqueous medium by activated carbon prepared from peanut shell: adsorption kinetics, equilibrium and thermodynamic studies. *Chemical Engineering Journal*, **184**, 238–247.

Amarasinghe, B.M.W.P.K. and Williams, R.A., 2007. Tea waste as a low cost adsorbent for the removal of Cu and Pb from wastewater. *Chemical Engineering Journal*, **132**, 299–309.

Amuda, O.S., Giwa, A.A., and Bello, I.A., 2007. Removal of heavy metal from industrial wastewater using modified activated coconut shell carbon. *Biochemical Engineering Journal*, **36**, 174–181.

Apak, R., Guclu, K., and Turgut, M.H., 1998. Modeling of copper (II), cadmium (II) and lead (II) adsorption on red mud. *Journal of Colloid and Interface Science*, **203**, 122–130.

Arbatov, A.A. Wastes as resources for sustainable development, *Regional Sustainable Development Review: Russia* (Related chapters), N. P. Laverov, Ed., Russian Academy of Sciences, Russia, http://www.eolss.net/Sample-Chapters/C16/E1-56-09.pdf.

Archer, M., and Baldwin, D., 2006. Composting seafood waste by windrow and in-vessel methods. *BioCycle*, **47**, 70–75.

Arvanitoyannis, I.S. and Kassaveti, A. 2007. Current and potential uses of composted olive oil waste. *International Journal of Food Science & Technology*, **42**, 281–295.

Babel, S. and Kurniawan, T.A., 2003. Low-cost adsorbants for heavy metals uptake from contaminated water: a review. *Journal of Hazardous Materials*, **B97**, 219–243.

Balakrishnan, M., Batra, V. S., Hargreaves, J. S. J., Monaghan, A., Pulford, I. D., Rico, J. L., and Sushil, S., 2009. Hydrogen production from methane in the presence of red mud—making mud magnetic. *Green Chemistry*, **11**, 42–47.

Balakrishnan, M., Batra, V., Hargreaves, J. S. J., and Pulford, I. D., 2011. Waste materials—catalytic opportunities: an overview of the applications of large scale waste materials as resources for catalytic applications. *Green Chemistry*, **13**, 16–24.

Balingar, V. C., Clark, R. B., Korcak, R. F., and Wright, R. J., 2011. Flue gas desulfurization product use on agricultural land. *Advances in Agronomy*, **111**, 51–86.

Belyaeva, O. N., Haynes, R. J., and Sturm, E. C., 2012. Chemical, physical and microbial properties and microbial diversity in manufactured soils produced from co-composting green waste and biosolids. *Waste Management*, **32**, 2248–2257.

Benito, M., Masaguer, A., De Antonio, R., and Moliner, A., 2005. Use of pruning waste compost as a component in soilless growing media. *Bioresource Technology*, **96**, 597–603.

Bernal, M. P., Alburquerque, J. A., and Moral, R., 2009. Composting of animal manures and chemical criteria for compost maturity assessment. A review. *Bioresource Technology*, **100**, 5444–5453.

Bhattacharya, A. K., Naiya, T. K., Mandal, S. N., and Das, S. K. 2008. Adsorption, kinetics and equilibrium studies on removal of Cr(VI) from aqueous solutions using different low-cost adsorbents. *Chemical Engineering Journal*, **137**, 529–541.

Bin Shawkataly, O., Jothiramalingam, R., Adam, F., Radhika, T., Tsao, T. M., and Wang, M. K., 2012. Ru-nanoparticle deposition on naturally available clay and rice husk biomass materials—benzene hydrogenation catalysis and synthetic strategies for green catalyst development, *Catalysis Science and Technology*, **2**, 538–546.

Boey, P.-L., Ganesan, S., Maniam, G. P., and Khairuddean, M., 2012. Catalysts derived from waste sources in the production of biodiesel using waste cooking oil, *Catalysis Today*, **190**, 117–121.

Bowden, L. I., Jarvis, A., Younger, J. P., and Johnson, K. L., 2009. Phosphorus removal from waste waters using basic oxygen steel slag. *Environmental Science & Technology*, **43**, 2476–2481.

Brown, P., Jefcoat, I. A., Parrish, D., Gill, S., and Graham, E., 2000. Evaluation of the adsorptive capacity of peanut hull pellets for heavy metals in solution. *Advances in Environmental Research*, **4**, 19–29.

Bulut, Y. and Zeki, T., 2007. Adsorption studies on ground shells of hazelnut and almond. *Journal of Hazardous Materials*, **149**, 35–41.

Cayuela, M. L., Sánchez-Monedero, M. A., and Roig, A., 2010. Two-phase olive mill waste composting: enhancement of the composting rate and compost quality by grape stalks addition. *Biodegradation*, **21**, 465–473.

Chambers, B., Nicholson, N., Smith, K., Pain, B., Cumby, T., and Scotford, I., 2001. Managing Livestock Manures. Booklet 1: Making Better Use of Livestock Manures on Arable Land; Booklet 2: Making Better Use of Livestock Manures on Grassland; Booklet 3: Spreading Systems for Slurries and Solid Manures. ADAS Gleadthorpe Research Centre, Notts, UK. http://archive.defra.gov.uk/foodfarm/landmanage/land-soil/nutrient/documents/manure/livemanure1.pdf.

Chen, C. and Wang, J., 2008. Removal of Pb^{2+}, Ag^+, Cs^+ and Sr^{2+} from aqueous solutions by brewery's waste biomass. *Journal of Hazardous Materials*, **151**, 65–70.

Chiang, Y. W., Ghyselbrecht, K., Santos, R. M., Meesschaert, B., and Matrens, J. A., 2012. Synthesis of zeolitic-type adsorbent materials from municipal solid waste incinerator bottom ash and its application in heavy metal adsorption. *Catalysis Today*, **190**, 23–30.

Cho, Y. B. and Seo, G., 2010. High activity of acid-treated quail eggshell catalysts in the transesterification of palm oil with methanol. *Bioresource Technology*, **101**, 8515–8519.

Conrad, K. and Hansen, H. C. B., 2007. Sorption of zinc and lead on coir. *Bioresource Technology*, **98**, 89–97.

Crini, G. and Badot, P.-M., 2008. Application of chitosan, a natural aminopolysaccharide, for dye removal from aqueous solutions by adsorption processes using batch studies: a review of recent literature. *Progress in Polymer Science*, **33**, 399–447.

Cui, J. and Zhang, L., 2008. Metallurgical recovery of metals from electronic waste: a review. *Journal of Hazardous Materials*, **158**, 228–256.

Danny, C. K., Cheung, C. W., Keith, K. H., and McKay, G., 2004. Sorption equilibria of metal ions on bone char. *Chemosphere*, **54**, 273–281.

De Resende, E. C., Gissane, C., Nicol, R., Heck, R. J., Guerreiro, C., Coelho, J. V., de Oliveira, L. C. A., Palmisano, P., Berruti, F., Briens, C., and Schlaf M., 2013. Synergistic co-processing of Red Mud waste from the Bayer Process and a crude untreated waste stream from bio-diesel production. *Green Chemistry*, **15**, 496–510.

Doğan, M., Abak, H., and Alkam, M., 2008. Biosorption of methylene blue from aqueous solution by hazelnut shells: equilibrium, parameters and isotherms. *Water, Air and Soil Pollution*, **192**, 141–153.

Feng, N., Guo, X., and Liang, S., 2009. Adsorption study of copper (II) by chemically modified orange peel. *Journal of Hazardous Materials*, **164**, 1286–1292.

Gadepalle, V. P., Ouki, S. K., van Herwijnen, R., and Hutchings, T., 2007. Immobilization of heavy metals in soil using natural and waste materials for vegetation establishment on contaminated sites. *Soil and Sediment Contamination*, **16**, 233–251.

Galea, B. 2010 Waste Regulation in India: An Overview. http://www.cppr.in/wp-content/uploads/2012/10/Waste-Regulation-in-India-An-Overview-Bernard.pdf, accessed May 10, 2013.

Gao, Y. and Xu, C., 2012. Synthesis of dimethyl carbonate over waste eggshell catalyst. *Catalysis Today*, **190**, 107–111.

Gelsomino, A., Abenavoli, M. R., Princi, G., Attina, E., Cacco, G., and Sorgana, A., 2010. Compost from fresh orange waste: a suitable substrate for nursery and field crops? *Compost Science and Utilization*, **18**, 201–210.

Gong, R. M., Ding, Y., Li, M., Yang, C., Liu, H., and Sun, Y., 2005. Utilization of powdered peanut hull as biosorbent for removal of anionic dyes from aqueous solution. *Dyes and Pigments*, **64**, 187–192.

Gu, H. N., Wang, N., and Liu, S. R., 2012. Radiological restrictions of using red mud as a building material additive. *Waste Management & Research*, **30**, 961–965.

Gupta, V. K., Gupta, M., and Sharma, S., 2001. Process development for the removal of lead and chromium from aqueous solutions using red mud—an aluminium industry waste. *Water Research*, **35**, 1125–1134.

Gupta, V. K. and Sharma, S., 2002. Removal of cadmium and zinc from aqueous solutions using red mud. *Environmental Science & Technology*, **36**, 3612–3617.

Gupta, V. K., Jain, C. K., Ali, I., Sharma, M., and Saini, K., 2003. Removal of cadmium and nickel from wastewater using bagasse fly ash—a sugar industry waste. *Water Research*, **37**, 4038–4044.

Gupta, V K. and Suhas, 2009. Application of low-cost adsorbants for dye removal—a review. *Journal of Environmental Management*, **90**, 2313–2342.

Gupta, V. K., Carrott, P. J. M., Ribeiro-Carrott, M. M. L., Suhas, 2009. Low-cost adsorbants: Growing approach to wastewater treatment—a review. *Critical Reviews in Environmental Science and Technology*, **39**, 783–842.

Hameed, B. H., 2009. Spent tea leaves: a new non-conventional and low-cost adsorbent for removal of basic dye from aqueous solutions. *Journal of Hazardous Materials*, **161**, 753–759.

Hargreaves, J. C., Adl, M. S., and Warman, P.R., 2008. A review of the use of composted municipal solid waste in agriculture. *Agriculture Ecosystems and Ecology*, **123**, 1–14.

Hargreaves, J. S. J. and Batra, V. S., 2012. Alternative sources of catalytic materials. *Catalysis Today*, 190.

Harrison, D. J., Bloodworth, A. J., Eyre, J. M., Macfarlane, M., Mitchell, C. J., Scott, P. W., and Steadman, E. J., 2002. Minerals from Waste: Project Summary Report, BGS Commissioned Report Cr/02/228N.

Hoornweg, D., Lam, P., and Chaudhry, M., 2005. Waste Management in China: Issues and Recommendations. Urban Development Working Papers East Asia Infrastructure Department, World Bank Working Paper No. 9. http://siteresources.worldbank.org/INTEAPREGTOPURBDEV/Resources/China-Waste-Management1.pdf, accessed on May 13, 2013.

Jala, S. and Goyal, D., 2006. Fly ash as a soil ameliorant for improving crop production—a review. *Biosource Technology*, **97**, 1136–1147.

Kamari, A., Pulford, I. D., and Hargreaves, J. S. J., 2011. Chitosan as a potential amendment to remediate metal contaminated soil—a characterization study. *Colloids and Interfaces B: Biointerfaces*, **82**, 71–80.

Karimi, E., Teixeira, I. F., Ribeiro, L. P., Gomez, A., Lago, R. M., Penner, G., Kycia, S. W., and Schlaf, M., 2012. Ketonization and deoxygenation of alkanoic acids and conversion of levulinic acid to hydrocarbons using a Red Mud bauxite mining waste as the catalyst. *Catalysis Today*, **190**, 73–88.

Karimi, E., Teixeira, I. F., Gomez, A., de Resende, E., Gissane, C., Leitch, J., Jollet, V., Aigner, I., Berruti, F., Briens, C., Fransham, P., Hoff, B., Schrier, N., Lago, R. M., Kycia, S. W., Heck, R., and Schlaf, M., 2013. Synergistic co-processing of an acidic hardwood derived pyrolysis bio-oil with alkaline Red Mud bauxite mining waste as a sacrificial upgrading catalyst. *Applied Catalysis B: Environmental*, doi 10.1016/j.apcatb.2013.02.007.

Karimian, N., Kalbasi, M., and Hajrasuliha, S., 2012. Effect of converter sludge, and its mixtures with organic matter, elemental sulfur and sulfuric acid on availability of iron, phosphorus and manganese of 3 calcareous soils from central Iran. *African Journal of Agricultural Research*, **7**, 568–576.

Kasongo, R. K., Verdoot, A., Kanyankagote, P., Baert, G., and Van Ranst, E., 2011. Coffee waste as an alternative fertilizer with soil improving properties for sandy soils in humid tropical environments. *Soil Use and Management*, **27**, 94–102.

Kastner, J. R., Miller, J., Geller, D. P., Locklin, J., Keith, L. H., and Johnson, T., 2012. Catalytic esterification of fatty acids using solid acid catalysts generated from biochar and activated carbon. *Catalysis Today*, **190**, 122–132.

Khemthong, P., Luadthong, C., Nualpaeng, W., Changsuwan, P., and Tongprem, P., 2012. Industrial eggshell wastes as the heterogeneous catalysts for microwave-assisted biodiesel production. *Catalysis Today*, **190**, 112–116.

Kim, E.-H., Yim, S.-B., Jung, H.-C., and Lee, E.-J., 2006. Hydroxyapatite crystallization from a highly concentrated phosphate solution using powdered converter slag as a seed material. *Journal of Hazardous Materials*, **B136**, 690–697.

Kim, J., Biswas, K., Jhon, K.-W., Jeong, S.-Y., and Ahn, W.-S., 2009. Synthesis of $AlPO_4$-5 and CrAPO-5 using aluminum dross. *Journal of Hazardous Materials*, **169**, 919–925.

Kumpiene, J., Lagerkvist, A., and Mauruce, C., 2008. Stabilization of As, Cr, Cu, Pb and Zn in soil using amendments—a review. *Waste Management*, **28**, 215–225.

Kuwahara, Y., Ohmichi, T., Kamegawa, T., Mori, K., and Yamashita H., 2010. A novel conversion process for waste slag: synthesis of a hydrotalcite-like compound and zeolite from blast furnace slag and evaluation of adsorption capacities. *Journal of Materials Chemistry*, **20**, 5052–5062.

Kuwahara, Y., Tsuji, K., Ohmichi, T., Kamegawa, T., Mori, K., and Yamashita, H., 2012. Waste-slag hydrocalumite and derivatives as heterogeneous base catalysts. *ChemSusChem*, **5**, 1523–1532.

Kuwahara, Y. and Yamashita, H., 2013. A new catalytic opportunity for waste materials: application of waste slag based catalyst in CO_2 fixation reaction. *Journal of CO$_2$ Utilization*, doi.org/10.1016/j.jcou.2013.03.001.

Kyzas, G. Z., Lazaridis, N. K., and Mitropoulos, A. C., 2012. Removal of dyes from aqueous solutions with untreated coffee residues as potential low-cost adsorbents: equilibrium, reuse and thermodynamic approach. *Chemical Engineering Journal*, **189–190**, 148–159.

Lafarge, 2008. From Waste to Resource: Creating a Sustainable Industrial System. Press kit. http://www.lafarge.com/06122008-press_themabook-industrial_ecology-uk.pdf.

Liu, Q., Xin, R., Li, C., Xu, C., and Yang, J., 2013. Application of red mud as a basic catalyst for biodiesel production. *Journal of Environmental Sciences*, **25**, 823–829.

Liu, S. Y., Gao, J., Yang, Y. J., Yang, Y. C., and Ye, Z. X., 2010. Adsorption intrinsic kinetics and isotherms of lead ions on steel slag. *Journal of Hazardous Materials*, **173**, 558–562.

Liu, Y., Naidu, R., and Ming, H., 2011. Red mud as an amendment for pollutants in solid and liquid phases. *Geoderma*, **163**, 1–12.

Lopez, F. A., Perez, C., Sainz, E., and Alonso, M., 1995. Adsorption of Pb^{2+} on blast furnace sludge. *Journal of Chemical Technology & Biotechnology*, **62**, 200–206.

Macquarrie, D. J. and Hardy, J. J. E., 2005. Applications of functionalized chitosan in catalysis. *Industrial and Engineering Chemistry Research*, **44**, 8499–8520.

Meunier, N., Laroulandie, J., Blais, J. F., and Tyagi, R. D., 2003. Cocoa shells for heavy metal removal from acidic solutions. *Bioresource Technology*, **90**, 255–263.

Murayama, N., Okajima, N., Yamaoka, S., Yamamoto, H., and Shibata, J., 2006. Hydrothermal synthesis of $AlPO_4$-5 type zeolitic materials by using aluminum dross as a raw material. *Journal of the European Ceramic Society*, **26**, 259–462.

Murthy, P. S. and Naidu, M.M., 2012. Sustainable management of coffee industry by-products and value addition: a review. *Resources, Conservation and Recycling*, **66**, 45–58.

Musyoka, N. M., Petrik, L. F., Hums, E., Baser, H., and Schweiger W., 2012. In situ ultrasonic monitoring of zeolite A crystallization from fly ash, *Catalysis Today*, **190**, 38–46.

Namasivayam, C. and Sureshkumar, M.V., 2008. Removal of chromium (VI) from water and wastewater using surfactant modified coconut coir pith as a biosorbant. *Bioresource Technology*, **99**, 2218–2225.

Nwachukwu, O. I. and Pulford, I. D., 2008. Comparative effectiveness of selected adsorbant materials as potential amendments for the remediation of lead-, copper- and zinc-contaminated soil. *Soil Use and Management*, **24**, 199–207.

Oguz, E., 2004. Removal of phosphate from aqueous solution with blast furnace slag. *Journal of Hazardous Materials*, **B114**, 131–137.

Oliveira, L. S., Franca, A. S., Alves, T. M., and Rocha, S. D., 2008a. Evaluation of untreated coffee husks as potential biosorbents for treatment of dye contaminated wastes. *Journal of Hazardous Materials*, **155**, 507–512.

Oliveira, W. E., Franca, A. S., Oliveira, L. S., and Rocha, S. D., 2008b. Untreated coffee husks as biosorbents for the removal of heavy metals from aqueous solutions. *Journal of Hazardous Materials*, **152**, 1073–1081.

Ormsby, R., Kastner, J. R., and Miller, J., 2012. Hemicellulose hydrolysis using solid acid catalysts generated from biochar. *Catalysis Today*, **190**, 89–88.

Ortiz, N., Pires, M. A. F., and Ressiani, J. C., 2001. Use of steel converter slag as nickel adsorber to wastewater treatment. *Waste Management*, **21**, 631–635.

Pan, X., Wang, J., and Zhang, D., 2009. Sorption of cobalt to bone char: kinetics, competitive sorption and mechanism. *Desalination*, **249**, 609–614.

Pande, G., Selvakumar S., Batra, V. S., Gardoll, O., and Lamonier, J-F., 2012. Unburned carbon from bagasse fly ash as a support for a VOC oxidation catalyst. *Catalysis Today*, **190**, 47–53.

Paredes, C., Bernal, M. P., Cegarra, J., and Roig, A., 2002. Bio-degradation of olive mill wastewater sludge by its co-composting with agricultural wastes. *Bioresource Technology*, **85**, 1–8.

Paredes, J. R., Ordonez, S., Vega, A., and Diez, F. V., 2004. Catalytic combustion of methane over red mud-based catalysts. *Applied Catalysis B- Environmental*, **47**, 37–45.

Park, D., Lim, S. R., Lee, H. W., and Park, J. M., 2008. Mechanism and kinetics of Cr(VI) reduction by waste slag generated from iron making industry. *Hydrometallurgy*, **93**, 72–75.

Park, J. E., Youn, H.-K., Yang, S.-T., and Ahn, W.-S., 2012. CO_2 capture and MWCNTs synthesis using mesoporous silica and zeolite 13X collectively prepared from incinerator bottom ash. *Catalysis Today*, **190**, 15–22.

Pehlivan, E., Altun, T., Cetin, S., and Bhanger, M. I., 2009. Lead sorption by waste biomass of hazelnut and almond shell. *Journal of Hazardous Materials*, **167**, 1203–1208.

Pontikes, Y. and Angelopoulos, G. N., 2009. Effect of firing atmosphere and soaking time on heavy clay ceramics with addition of Bayer's process bauxite residue. *Advances in Applied Ceramics*, **108**, 50–56.

Pradhan, J., Das, S. N., and Thakur, R. S., 1999. Adsorption of hexavalent chromium from aqueous solution by using activated red mud. *Journal of Colloid and Interface Science* **217**, 137–141.

Pulford, I. D., Hargreaves, J. S. J., Ďurišová, J., Kramulova, B., Girard, C., Balakrishnan, M., Batra, V., and Rico, J. L., 2012. Carbonised red mud—a new water treatment product made from a waste material. *Journal of Environmental Management*, **100**, 59–64.

Rafiee, E., Khodayari, M., Kahirizi, M., and Tayebee, R., 2012. $H_5CoW_{12}O_{40}$ supported on nano silica from rice husk ask: a green bifunctional catalyst for the reaction of alcohols with cyclic and acyclic 1,3-dicarbonyl compounds. *Journal of Molecular Catalysis: Chemical*, **358**, 121–128.

Rattanashotinut, C., Thairit, P., and Tangchirapat, W., 2013. Use of calcium carbide residue and bagasse ash mixtures as new cemenitious material in concrete. *Materials and Design*, **46**, 106–111.

Rios R., C. A., Oviedo V., J. A., Henao M., J. A., and Macias L., M. A., 2012. A NaY zeolite synthesized from Colombian industrial coal by-products; potential catalytic applications. *Catalysis Today*, **190**, 61–67.

Rode, C. V., Mane, R. B., Potdar, A. S., Patil, P. B., Niphadkar, P. S., and Joshi, P. N., 2012. Copper modified waste fly ash as a promising catalyst for glycerol hydrogenolysis. *Catalysis Today*, **190**, 31–37.

Roig, A., Cayuela, M. L., and Sánchez-Monedero, M. A., 2006. An overview on olive mill wastes and their valorisation methods. *Waste Management*, **26**, 960–969.

Santona, L., Castaldi, P., and Melis, P., 2006. Evaluation of the interaction mechanisms between red muds and heavy metals. *Journal of Hazardous Materials*, **132**, 324–329.

Saptura, E., Muhammad, S., Sun, H., Ang, H. M., Tadé M. O., and Wang, S., 2012. Red mud and fly ash supported Co catalysts for phenol oxidation. *Catalysis Today*, **190**, 68–72.

Schiewer, S. and Patel, S. B., 2008. Pectin-rich fruit wastes as biosorbants for heavy metal removal: equilibrium and kinetics. *Bioresource Technology*, **99**, 1896–1903.

Schlummer, M., Maurer, A., Leitner, T., and Spruzina, W., 2006. Report: recycling of flame-retarded plastics from waste electric and electronic equipment (WEEE). *Waste Management & Research*, **24**, 573–583.

Schussler, S., Blaubach, N., Stolle, A. S., Cravatto, G., and Ondruschka, B., 2012. Application of a cross-linked Pd-chitosan catalyst in liquid-phase-hydrogenation using molecular hydrogen. *Applied Catalysis A – General*, **445**, 231–238.

Šćiban, M., Klašnja, M., and Škrbić, B., 2006. Modified hardwood sawdust as adsorbant of heavy metal ions from water. *Wood Science and Technology*, **40**, 217–227.

Šćiban, M., Klašnja, M., and Škrbić, B., 2008. Adsorption of copper ions from water by modified agricultural by-products. *Desalination*, **229**, 170–180.

Shah, B., Mistry, C., and Shah, A., 2013. Seizure modelling of Pb(II) and Cd(II) from aqueous solution using chemically modified sugarcane bagasse fly ash: isotherms, kinetics and column studies. *Environmental Science Pollution Research*, **20**, 2193–2209.

Shah, B. A., Tailor, R. V., and Shah, A. V., 2012. Zeolitic bagasse fly ash as a low cost sorbent for the sequestration of p-nitrophenol: equilibrium, kinetics and column studies. *Environmental Science Pollution Research*, **19**, 1171–1186.

Shukla, A., Zhang, Y.-H., Dubey, P., Margrave, J. L., and Shukla, S. S., 2002. The role of sawdust in the removal of unwanted materials from water. *Journal of Hazardous Materials*, **95**, 137–152.

Siles López, J. A., Li, Q., and Thompson, I. P., 2010. Biorefinery of waste orange peel. *Critical Reviews in Biotechnology*, **30**, 63–69.

Singh, S. R. and Singh, A. P., 2012. Treatment of water containing chromium (VI) using rice husk carbon as a new low cost adsorbent. *International Journal of Environmental Research*, **6**, 917–924.

Sivaraj, R., Namasivayam, C., and Kadirvelu, K., 2001. Orange peel as an adsorbant in the removal of Acid violet 17 (acid dye) from aqueous solutions. *Waste Management*, **21**, 105–110.

Sivrikaya, S., Albayrak, S., Imamoglu, M., Gundogdu, A., Duran, C., and Yildiz, H., 2012. Dehydrated hazelnut husk carbon: a novel sorbent for the removal of Ni(II) ions from aqueous solution. *Desalination and Water Technology*, **50**, 2–13.

Smith, S. R., 1996. *Agricultural Recycling of Sewage Sludge and the Environment*. CAB International, Wallingford, UK.

Song, C. and Jiang, Y., 2012. The utilization of industrial solid waste in metallurgical industry, World Automation Congress (WAC) June 24–28, 2012, Puerto Vallarta, Mexico.

Sushil, S. and Batra, V. S., 2008. Catalytic applications of red mud, an aluminium industry waste: a review. *Applied Catalysis B: Environmental*, **81**, 64–67.

Sushil, S., Alabdulrahman, A. M., Balakrishnan, M., Batra, V.S., Blackley, R. A., Clapp, J., Hargreaves, J. S. J., Monaghan, A., Pulford, I. D., Rico, J. L., and Zhou, W., 2010. Carbon deposition and phase transformations in red mud on exposure to methane. *Journal of Hazardous Materials*, **180**, 409–418.

Sushil, S. and Batra, V. S., 2012. Modification of red mud by acid treatment and its application for CO removal, *Journal of Hazardous Materials*, **203–204**, 264–273.

Tarley, C. R. T. and Arruda, M. A. Z., 2004. Biosorption of heavy metals using rice milling by-products. Characterisation and application for removal of metals from aqueous effluents. *Chemosphere*, **54**, 987–995.

Theivarasu, C. and Mylasamy, S., 2012. Adsorption studies of Acid Blue-92 from aqueous solution by activated carbon obtained from agricultural industrial waste – Cocoa (*Theobroma cacoa*) shell. *Asian Journal of Chemistry*, **24**, 2187–2190.

Tsakiridis, P.E., Agatzini-Leonardou, S., and Oustadakis, P., 2004. Red mud addition in the raw meal for the production of Portland cement clinker. *Journal of Hazardous Materials*, **116**, 103–110.

Tuttobene, R., Avola, G., Gresta, F., and Abbate, V., 2009. Industrial orange waste as organic fertiliser for durum wheat. *Agronomy for Sustainable Development*, **29**, 557–563.

UNCRD, 2011. Country analysis paper—China, Third Meeting of the Regional 3R Forum in Asia, "Technology Transfer for Promoting the 3Rs – Adapting, Implementing, and Scaling Up Appropriate Technologies," Singapore, October 5–7, 2011. http://www.uncrd.or.jp/env/spc/docs/3rd_3r/Country_Analysis_Paper_China.pdf, accessed on May 10, 2013.

UNEP, 2009. *Critical Metals for Future Sustainable Technologies and Their Recycling Potential*. Report by Öko-Institut e.V., UNEP DTIE, Sustainable Consumption Branch, Paris (www.unep.fr).

van Herwijnen, R., Laverye, T., Poole, J., Hodson, M.E., and Hutchings, T.R., 2007. The effect of organic materials on the mobility and toxicity of metals in contaminated soils. *Applied Geochemistry*, **22**, 2422–2434.

Vadvelan, V. and Kumar, K.V., 2005. Equilibrium, kinetics, mechanisms, and process design for the sorption of methylene blue onto rice husk. *Journal of Colloid and Interface Science*, **286**, 90–100.

Viriya-empikul, N., Krasae, P., Puttasawat, B., Yoosuk, B., Chollacoop, N., and Faungnawakji, K. 2010. Waste shells of mollusk and egg as biodiesel production catalysts, *Bioresource Technology*, **101**, 3765–3767.

Wang, S., Ang, H.M., and Tadé, M.O., 2008. Novel applications of red mud as coagulant, adsorbent and catalyst for environmentally benign processes. *Chemosphere*, **72**, 1621–1635.

Wang, X.J., 1998. Waste Reuse: Legislation and Enforcement in China, Fifth International Conference on Environmental Compliance and Enforcement, November 16–20, 1998 Monterey, CA.

Wei, Z., Xu, C., and Li, B., 2009. Application of waste eggshell as low-cost solid catalyst for biodiesel production, *Bioresource Technology*, **100**, 2883–2885.

Wells, D.E., Sibley, J.L., Gillham, C.H., and Dozier, W.A., 2012. Use of composted and fresh spent tea grinds as a potential greenhouse substrate component. *Compost Science and Utilization*, **20**, 181–184.

Wilson, J.A., Pulford, I.D., and Thomas, S., 2003. Sorption of Cu and Zn by bone charcoal. *Environmental Geochemistry and Health*, **25**, 51–56.

Wu, Y., Jiang, L., Wen, Y.-J., Zhou, J.-X., and Feng, S., 2012. Biosorption of Basic Violet 5BN and Basic Green by waste brewery's yeast from single and multicomponent systems. *Environmental Science and Pollution Research*, **19**, 510–521.

Xue, Y., Hou, H., and Zhu, S., 2009. Competitive adsorption of copper(II), cadmium(II), lead(II) and zinc(II) onto basic oxygen furnace slag. *Journal of Hazardous Materials*, **162**, 391–401.

Zeng, M. F., Zhang, X., Qi, C. Z., and Zhang, X.M., 2012. Microstructure-stability relations studies of porous chitosan microspheres supported palladium catalysts, *International Journal of Biological Macromolecules*, **51**, 730–737.

Zhang, Y., Liu, W., Xu, M., Zheng, F., and Zhao, M., 2010. Study of the mechanisms of Cu^{2+} biosorption by ethanol/caustic-pretreated baker's yeast biomass. *Journal of Hazardous Materials*, **178**, 1085–1093.

Zhou, Y.-F. and Haynes, R.J., 2011. A comparison of inorganic solid wastes as adsorbants of heavy metal cations in aqueous solution and their capacity for desorption and regeneration. *Water, Air, Soil Pollution*, **218**, 457–470.

Zhu, B., Fan, T., and Zhang, D., 2008. Adsorption of copper ions from aqueous solution by citric acid modified soybean straw. *Journal of Hazardous Materials*, **153**, 300–308.

2 Waste from Metal Processing Industries

M. Balakrishnan, V.S. Batra, and J.S.J Hargreaves

CONTENTS

Metal ores are complex mixtures of compounds and several steps are used to obtain the respective metals from them. Pyrometallurgical, hydrometallurgical, or electrolytic methods are used to extract the metals from the ore and to refine them depending on the end-use requirements. These steps involve generation of wastes in large quantities whose properties and hazardous nature vary considerably. With increasing environmental awareness and rising disposal costs, these wastes are being seen as raw materials for other applications. In this chapter, the status of use of wastes generated from different metal industries is discussed and the trends in research in this area are presented. In case of wastes such as slag where utilization levels are high, the current common applications are covered. However, for wastes like red mud, the focus is more on literature studies since the utilization levels are still low and many exploratory studies on their use are underway.

2.1   IRON AND STEEL

Iron is produced in a blast furnace (BF) where iron ore, coke, and limestone are charged from the top. In a series of reactions, the iron oxide is reduced to iron by the reducing gases from the coke and the limestone flux reacts with the impurities mainly comprised of silica and alumina to form low melting slag. The molten metal is then taken for steel making. In a basic oxygen furnace (BOF) (or basic oxygen converter), the molten iron is subjected to the addition of flux and blowing of oxygen. This oxidizes the impurities like silicon, which react with the flux to form slag. Steel is also made in an electric arc furnace (EAF) where fluxes are added which react with impurities to form slag. In addition to slag, there are also other wastes generated as part of the flue gas cleaning system, furnace feed preparation, etc. Figure 2.1 shows a flow chart for iron and steel making with the waste generation points.

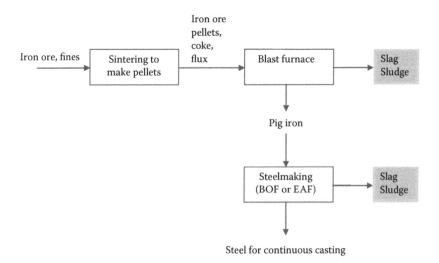

FIGURE 2.1   Iron and steel making process showing the wastes generated.

The world production of iron and steel is shown in Table 2.1. Due to the differences in the type of ore, process steps used, local legislation, etc., the nature of the wastes generated, their management and utilization steps, and research focus varies. These are elaborated in this section.

2.1.1 SLAG

The amount of slag produced in the blast furnace stage is around 300 to 400 kg per tonne of iron (Kumar et al., 2006); it has also been reported that the slag generation per tonne of iron can vary from 220 to 370 kg for an ore with 60 to 65% iron content to 1 tonne in case of a poor grade ore (Kalyoncu, 2000). Steel making generates 20 wt% slag and once the steel present in the slag is recovered, it comes down to 10 to 15% of the weight of steel produced (Kalyoncu, 2000). The amount of slag generated and utilized in different countries is shown in Table 2.2.

The composition of the slag varies depending on the ore and flux used and the process employed. Table 2.3 summarizes the compositions of different iron and steel making slags from different reports in literature. In addition to the composition, in case of blast furnace slag, the cooling method also influences its properties and hence the applications. The slag can be air cooled, cooled in water or steam to form expanded slag, or rapidly cooled in water to form granulated slag.

2.1.1.1 Slag Utilization

In the European Union (EU), almost 87% of the slag is used for building applications. There are many standards and regulations in the EU concerning use of slag and discussions are underway to classify slag as waste, non-waste, or by-product (Bialuch et al., 2011). In Germany, 90% of BF slag steel slag is utilized (Mudersbach et al., 2011). In the United States, it is estimated that of the approximately 20 million tonnes generated in 2011, 17 million tonnes were sold (van Oss, 2012). In Japan, almost all the slag was used with almost 69% of blast furnace slag being used for cement application (Nippon Slag Association, 2012a) and almost 60% of steel slag being used for road applications (Nippon Slag Association, 2012b).

The utilization of slag has been the subject of review in several studies (Das et al., 2007; Dippenaar, 2004; Mihok et al., 2006; Reuter et al., 2004; Singh and Raju, 2011; Vlcek et al., 2012). In a 2002 review of utilization of steel slag in Saudi Arabia, the common uses were reported to be as asphalt and concrete aggregates with nearly 67% of steel slag being used in the road construction industry (Khan et al., 2002). Many reviews have focused on specific applications. Dziarmagowski (2005) examined the use of converter slag as a charge in blast furnaces or for reduction in electric arc furnaces where the metallic iron component in the slag is utilized. The non-metallic portion could be used for Portland cement (Dziarmagowski, 2005). The use of blast furnace slag for geopolymers and the mechanism of formation have been reviewed (Davidovits, 2011). As part of a review on utilization of wastes from metallurgical and allied industries, the use of granulated blast furnace slag in blended cements and tiles is described (Kumar et al., 2006). Reviews on use of steel slag for road construction and hydraulic structures have also been published (Motz and Geiseler, 2001; Sofilic et al., 2012; Wang and Emery, 2004). The recovery of metals from slag

TABLE 2.1
(a) Pig Iron and (b) Crude Steel Production in 2011 (thousand tonnes)

(a)

Country	Pig Iron Production
China	629693
Japan	81028
Russia	48120
South Korea	42218
India	38900
Brazil	33243
United States	30233
Ukraine	28867
Germany	27795
Italy	9824
France	9698
Turkey	8173
Canada	7520
Australia	5265
South Africa	4765

(b)

Country	Crude Steel Production
China	683265
Japan	107595
United States	86247
India	72200
Russia	68743
South Korea	68471
Germany	44288
Ukraine	35332
Brazil	35162
Turkey	34103
Italy	26662
Taiwan	22660
France	15777
Spain	15591
Canada	13090
Iran	13040
United Kingdom	9481
Poland	8794
South Africa	6650
Egypt	6486
Australia	6404

Source: World Steel Association, 2011.

TABLE 2.2
Slag Generation in Different Countries (million tonnes)

Country	Slag Generated (year)	Slag Utilized	Ref.
United States	16–21 (2011)	17.00	van Oss, 2012
Australasia	2.84 (2010)	2.67	Australasia Slag Association, 2010
Europe	45.3 (2010)	47.90	The European Slag Association, 2012
Japan	38.37 (2011)	40.86	Nippon Slag Association, 2012a, 2012b

has been reviewed (Shen and Forssberg, 2003). Two reviews by Shi focus on the application of steel and other metallurgical slags for cement applications (Shi and Qian, 2000; Shi, 2004). In the review by Vlcek et al. (2012), use of alkali-activated slag as a binder for cement was examined. Wu and Zongshu (2005) examined the use of iron and steel slags in agriculture. A review of steel slag generation and its characteristics with a focus on chemical and mineralogical data has been carried out (Yildirim and Prezzi, 2011).

While slag utilization amounts are high in most developed countries, there are also research studies for other improvements in properties, new applications, etc. For some applications, the stability of the slag and its environmental impact in terms of leaching are important. Slags from ladle furnace, BOF, and EAF were subjected to various cooling conditions to study the effect on volume stability and leaching; it was observed that these did not prevent leaching (Tossaveinen et al., 2007). To understand the volume instability of ladle slag, a detailed characterization was undertaken and it was concluded that lime, which causes the instability, could be present in different forms (nodules or micro inclusions); the nature of lime also influences the extent of instability (Waligora et al., 2010).

2.1.1.2   Hot Processing

An area that is receiving increasing attention is hot stage processing of slag where modifications to the slag are made when it is in the molten stage. Modifications in terms of additives or cooling procedure can be made to improve the utilization of the slag including reducing the leaching potential. This has been reviewed with different case studies such as borate or silica additions to prevent slag disintegration due to volume changes and additions of iron oxide and alumina (from bauxite, for example) to reduce chromium leaching (Durinck et al., 2008). Developments in hot stage processing in China have been reviewed. The options include heat exchange during granulation to recover heat followed by stabilization of lime with steam and CO_2 and finally magnetic separation of metals. In another scheme, slag is homogenized by addition of waste clay bricks, glass cullet, etc., which react with free lime to form cementitious material; this is followed by water quenching and granulation and magnetic separation (Li and Ni, 2011). Borate-based stabilizer has been added to inhibit the beta to gamma transformation in Ca_2SiO_4 (Engstrom et al., 2011). Other examples of slag modification are addition of lime to improve basicity ratio of water granulated BF slag, which led to increase in compression strength of mortar; incorporation of silica-containing material to bind free lime in BOF slag, which could cause volume

TABLE 2.3
Typical Slag Compositions (wt%)

Type of slag	CaO	SiO$_2$	Al$_2$O$_3$	MgO	Fe$_2$O$_3$	MnO	K$_2$O	Na$_2$O	TiO$_2$	S	BaO	Ref.
BF slag, Turkey	36.80	36.00	14.84	7.10			0.81	0.30				Camci et al., 2002
Granulated BF slag, Taiwan	41.67	43.39	14.47	6.49			0.36	0.22	0.53			Cheng and Chiu, 2003
Granulated BF slag, Egypt	29.83	41.46	11.36	1.35	0.63	4.95	0.33	0.61	0.48		6.35	El-Mahllawy, 2008
BF slag, China	41.78	40.55	7.60	6.16	0.85							Liu et al., 2007
BF slag, Iran	37.20	36.20	8.00	10.30		0.90			4.70	1.50		Monshi and Asgarani, 1999
BF slag, Japan	42.99	33.52	13.77	6.28		0.27						Kashiwaya et al., 2010
BOF steel slag, China	44.73	14.53	4.60	8.46						0.05		Li and Ni, 2011
BOF steel slag, Turkey	38.62	19.28	2.71	8.05	22.61	7.52	0.13	0.28	0.52	0.28		Alanyali et al., 2006
Steel slag, Saudi Arabia	24.48	14.20	8.12	10.76	34.89	3.87			1.14			Khater, 2002
Steel slag, South Africa	36.28	16.90	2.86	10.33		0.78				0.12		Mulopo et al., 2012
EAF slag, Greece	32.50	18.10	13.30	2.53	26.30	3.94		0.13	0.47		0.14	Iacobescu et al., 2011
EAF slag, India	22.80	20.30	7.30	8.00	42.40		0.82	0.63	0.32			Sarkar et al., 2010

instability in road applications; and binding of chromium in the form of spinel in stainless steel slag (Mudersbach et al., 2011). Use of boron-containing waste for such a modification has also been described (Pontikes et al., 2011). To prevent volume instability due to hydration and carbonation of free lime, hot stage carbonation of BOF slag has been examined to form calcium carbonate (Santos et al., 2012).

2.1.1.3 Recovery of Iron and Other Compounds

Separation of magnetic portions of a BOF slag from a steel plant has been studied and optimum design and operating conditions for a drum magnetic separator have been discussed. The recovered magnetic portions of the slag have potential to be recycled as raw material in a blast furnace or as scrap in steel making (Alanyali et al., 2006). Recovery of manganese and chromium from steel slag has been studied with a focus on removal of phosphorus so that the recovered material can be used in alloying the steel. This was achieved by a sulfurization step, which leads to the formation of an iron and manganese sulfide matte and a slag where the phosphorus is separated. The matte was then further processed to recover low phosphorus ferro manganese alloy (Kitamura et al., 2011). The effect of recycling demetallized EAF slag back in the process was examined. It was found that there was no negative impact on the composition of the steel produced but the amount of lime added in the charge could be reduced (Mihok et al., 2004). A scheme for recovering iron and phosphorus has been proposed based on experiments with synthesized slag and Fe-P-C alloy. Microwave carbothermal reduction of the slag leads to formation of metallic iron phase containing phosphorus and carbon. Subsequently, this can be treated with a potassium or sodium carbonate flux, which removes the P from the melt forming phosphate. The phosphate has potential to be used as fertilizer while the metal can be recycled back in the process (Morita et al., 2002). Calcium carbonate has been recovered from steel slag by extraction with hydrochloric acid, and carbonation of the solution with CO_2 after neutralization, which led to the precipitation of calcium carbonate. The recovered calcium carbonate could neutralize acid mine drainage; however, the presence of trace elements, which could be toxic in the leachate, needs to be addressed (Mulopo et al., 2012). A process of recovering iron from converter slag using magnetic separation has been reported. After recovering the magnetic iron particles, the non-magnetic ferric oxide in slag was converted to magnetic iron by reduction below 1000°C with hydrogen and carbon monoxide followed by magnetic separation. The non-magnetic portion after removal of excess calcium oxide was suitable for cement applications (Gao et al., 2011).

2.1.1.4 Cement and Concrete

BF slag and steel slags from BOF and EAF have been used along with water glass for cementless concrete, which had good compression strength (Baricova et al., 2012). EAF slag has been studied for making belite cement (comprised mainly of Ca_2SiO_4), which requires less energy to produce compared to ordinary Portland cement. The cements were prepared successfully; however, the early strength was less due to the presence of high iron content from the slag (Iacobescu et al., 2011). The use of both BF slag and steel slag for Portland cement has been reported (Monshi and Asgarani, 1999). To use stainless steel slag as a hydraulic binder, mechanical activation by milling was carried out. This led to increase in reactivity; however, the strength developed when mixed with water was less than that of ordinary Portland cement (Kriskova et al., 2012). Fly ash slag mixtures have been subjected to alkali activation and the cement pastes obtained have been characterized in detail (Puertas and Jimenez, 2003). Ferromolybdenum slag has been used as an aggregate in concrete blocks. There was no deterioration in the properties and leaching of heavy metals did not exceed the regulated values (Boehme and Hende, 2011). The use of blast furnace slag in geopolymers has been reviewed with particular emphasis on using conditions that avoid free alkali in the system (Davidovits, 2011). Granulated blast furnace slag has been used in the synthesis of geopolymers. The mechanical and fire resistance properties were satisfactory with potential for construction and other engineering applications (Cheng and Chiu, 2003).

  EAF slag has been used as an aggregate in bituminous paving mixtures and showed suitable properties, with leaching below the limits set by Italian standards (Sorlini et al., 2012). Steel slag has been used in stone mastic asphalt and showed better high temperature properties compared to basalt (Wu et al., 2007). The use of slag in road construction has been reviewed. Slag has been used in both asphalt mixtures and in unbound base courses (Barisic et al., 2010).

2.1.1.5 Ceramics (Bricks, Tiles)

Granulated blast furnace slag has been combined with another waste—quarry fines—to make fired acid-resistant bricks. Different compositions and firing temperatures were studied and optimum composition based on the water absorption, acid weight loss, and compression strength were identified with 30% blast furnace slag (El-Mahllawy, 2008). Slag from stainless steel EAF has been used in making bricks along with clay. It was observed that bricks with less than 10% slag addition were suitable (Shih et al., 2004). Water granulated BF slag has been used to make porous sound-absorbing blocks by compression molding and high-temperature sintering at 1100°C and showed promise for this application (Peng et al., 2011). EAF steel a slag has been used along with clay, quartz, and feldspar to prepare vitreous tiles, which showed good strength depending on the composition and firing temperature (Sarkar et al., 2010).

  BF slag has been used to prepare glass ceramics by removal of magnetic metal from crushed slag, melting of remaining glass, and crystallization. However, the resulting glass ceramic did not have strength comparable to that of glass ceramics used in the construction industry (Fredericci et al., 2000). Glass ceramics have been prepared from steel slag along with addition of sand, clay, dolomite, and limestone followed by melting and annealing. Up to 57 wt% of steel slag could be used in the compositions (Khater, 2002). The application of ferrous tailings and slag in glass ceramics, particularly in China, has been discussed (Liu et al., 2007). Up to 3 wt% steel solid waste has been added to kaolinitic clay for making bricks. It was observed that the mechanical properties and open porosity showed potential for building applications and the leaching of heavy metals was within limits (Oliveira and Holanda, 2004). Steel slag mixed with kaolinitic clay has been prepared for brick and roofing tiles. With up to 5 wt% addition, there was no deterioration in mechanical properties with the added advantage of lower linear shrinkage on firing (Vieira et al., 2006).

2.1.1.6   Adsorbent

Blast furnace slag has been used as an adsorbent for phosphorus removal from wastewater and, depending on the adsorption conditions, up to 99% removal could be obtained (Oguz, 2004). Slag from BOF has been used as an adsorbent for removal of phosphorus from wastewater and it was observed that the phosphorus phases formed on the surface were stable (Bowden et al., 2009). Field data from a filter media made of steel slag used in the polishing step as part of a wastewater treatment system has shown that phosphorus removal also took place. They were effective for about 5 years after which the performance declined (Shilton et al., 2006). Steel slag has been used for phosphate removal and was found to be effective for this application (Xiong et al., 2008).

EAF slag from stainless steel has been used for absorption of CO_2 using wet grinding and dry grinding in the presence of CO_2. The absorption takes place by carbonation and formation of calcium carbonate from the reaction with calcium oxide present in the slag. The study observed that wet grinding was more effective compared to dry grinding (Hisyamudin et al., 2009). BF slag has been used to remove Cr (VI) from wastewater, which gets reduced to less toxic Cr (III) due to presence of iron in the zerovalent or divalent form in the slag (Park et al., 2008). Steel slag has been used for removal of lead and, based on the particle size, up to 97% removal was obtained (Liu et al., 2010). BOF slag has been tested for removal of cadmium, copper, lead, and zinc in single component and multi-component systems (Xue et al., 2009).

2.1.1.7   Others

In an attempt to use amorphous BF slag as phase change material, the latent heat of crystallization was measured and simulation was done to test the use of this latent heat for reducing the temperature of waste gas to a more useable temperature (Kashiwaya et al., 2010). Hydrothermal treatment of slag was performed by hot pressing a slag water mixture in an autoclave in a specially designed hydrothermal hot pressing apparatus. This material was tested as a lubricant in the steel rolling process (Sato et al., 2008). In titanium-containing BF slag as in China, the precipitation of titanium-rich perovskite and its growth kinetics have been studied (Wang et al., 2008). A procedure for making ferroalloy doped with P, Mn, and Cr has been described. The slag coke mixture was heated to reduce the metal oxides and the reduced metal was separated by magnetic separation. This was used as a precursor for making doped $FePO_4$ by reacting with phosphoric acid and hydrogen peroxide, which could then be used for making $LiFePO_4$ for lithium batteries (Wu et al., 2011). A study examined the dissolution behavior of synthesized slag in seawater with a view to use slag for rehabilitation of coastal regions because the dissolution of ferrous iron from slag in seawater can promote sea plant growth (Zhang et al., 2011). BF slag has been used as a starting material for preparing a hydrocalumite base catalyst by acid dissolution and precipitation. These were used for transesterification of esters for biofuel applications with promising results (Kuwahara et al., 2012a). The hydrocalumite catalyst was also active in other reactions such as Knoevenagel condensation, alkylaromatics oxidation, and cycloaddition of epoxide with CO_2 (Kuwahara, 2012b). BF slag has been used as a catalyst for generating hydrogen from biogas and was found to be active at temperatures of 900 to 1000°C. It was therefore postulated that the waste heat from the slag could be utilized for this reaction (Purwanto and Akiyama, 2006).

2.1.2 SLUDGE

The BF and BOF have flue gas cleaning systems and the dust is collected in the form of slurry. The water from the slurry is pressed out and the remaining solid forms the sludge. The amount of BF dust is between 8 to 12 kg/tonne of pig iron (Veres et al., 2011) while in a BOF, the sludge generated is around 27 kg/tonne of crude steel on dry basis (Singh and Raju, 2011). It is estimated that 30 million tonnes of dust are released globally each year from iron and steel making (Kumar and Liu, 2011). The composition of sludge is shown in Table 2.4.

2.1.2.1 Recycling of Sludge Back in Process

Many studies have examined the use of sludge. Some of the studies have examined the recycling of converter sludge back in the process. A recent review has examined the practice of sludge utilization particularly from the point of view of recycling in the sinter plant for blast furnaces feed (Singh and Raju, 2011). Sludge from BF and BOF were mixed with other solid wastes (dust, mill scale, etc.) to form pellets with 57.2% iron, which were assessed as potential feeds for blast furnaces (Camci et al., 2002). A similar study on the preparation of cold-bonded pellets using BOF sludge, BF dust, and oily mill sludge has been reported. The composition and preparation conditions were optimized to obtain pellets suitable as converter feed (Su et al., 2004). A U.S. patent describes a process for separating metallic iron from sludge and agglomerating it to make it suitable for recycling in steel making (Gomez and Santoz, 2012). The sintering of converter sludge and dust in the presence of coke and limestone was studied to understand the removal of zinc and lead during this process. This causes the formation of sinter with low zinc and lead amounts, making it suitable for recycling; further, the lead and zinc vapors can potentially be concentrated and utilized for extraction of these metals

TABLE 2.4
Typical Dust and Sludge Compositions (wt%)

Type of Waste	CaO	SiO$_2$	Al$_2$O$_3$	MgO	Fe total	Zn	Ref.
BF dust	3.52	5.78	2.41	0.74	61.00	0.07	Camci et al., 2002
BF sludge	3.66	5.56	1.94	1.21	81	0.31	Camci et al., 2002
BOF dust	19.40	2.88	0.20	1.50	20.00	0.03	Camci et al., 2002
BOF sludge	7.85	3.77	1.04	0.39	60.00	0.37	Camci et al., 2002
BF dust	5.16	4.98	1.93	1.06	46.7 (FeO)	0.23	Su et al., 2004
BOF coarse sludge	17.41	1.38	0.20	3.71	46.7 (FeO)	0.05	Su et al., 2004
BF sludge	4.28	7.02	1.74	1.87	41.44	0.77	Veres et al., 2011
BOF dust	8.41	3.98	0.85	1.86	71.48	0.01	Wang et al., 2010
BOF sludge	11.45	3.48	0.85	3.30	47.92	0.16	Wang et al., 2010

(Legemza, 2004). The zinc present in the converter sludge and dust has been examined for potential desulfurization of steel on the laboratory scale (Kumar and Liu, 2011). Dolomite sludge mix was prepared by mixing the sludge with fines in flue gas resulting from calcinations/sintering of dolomite; when added to iron ore fines, this mix prevented dusting and free flowing material for blast furnaces could be obtained (Agrawal and Pandey, 2005).

2.1.2.2 Metal Recovery

BF sludge has been subjected to a hydrometallurgical route for zinc extraction. The sludge was leached with sulfuric acid under different operating conditions and it was observed that microwave leaching yielded quicker and higher recovery of zinc compared to conventional leaching (Veres et al., 2011). BOF sludge was mixed with small amounts of coal powder and heated to 1330°C, which led to reduction and formation of iron nuggets that could be separated from the slag by screening (Wang et al., 2010).

2.1.2.3 Other Applications

Converter sludge has been tested as a soil amender for calcareous soil in combination with organic matter and sulfur; it was found to improve the iron availability in the soil (Karimian et al., 2012). Converter sludge has also been used as an adsorbent to remove Ni from wastewater (Ortiz et al., 2001). Converter sludge and several other low-cost adsorbents were tested for Cr(VI) removal from wastewater. It was observed that converter sludge had removal of more than 99% under certain experimental conditions (Bhattacharya et al., 2008). Converter sludge has been used as a seed for the crystallization of hydroxyapatite from phosphorus-containing wastewater as a means to recover phosphorus (Kim et al., 2006). Converter sludge has been used to prepare doped $FePO_4$ by first leaching the slag with sulfuric acid to obtain the dissolved metal sulfates, which are then reacted with phosphoric acid and then precipitated and calcined to obtain the doped $FePO_4$ for lithium battery applications (Wu et al., 2012). BF sludge showed potential for lead adsorption with a capacity of 80 mg lead per gram of sludge (Lopez et al., 1995).

2.2 ALUMINIUM

Aluminium metal is characterized by light weight, high strength, easy workability, good heat and electrical conductivity, and good corrosion resistance. Consequently, it finds use in a variety of applications in packaging, aerospace, buildings, and consumer goods. There are currently two routes of aluminium production, namely from the ore (primary production) and from recycling aluminium products (secondary production). Bauxite is the main ore containing a 45 to 60% mixture of aluminium hydroxides, namely gibbsite $Al(OH)_3$, boehmite γ-AlO(OH), and diaspore α-AlO(OH). The other components in the ore are hydroxides/oxides of iron (goethite FeO(OH) and hematite Fe_2O_3), clay, and anatase (TiO_2). Bauxite is refined to produce alumina (Al_2O_3); the alumina is further smelted electrolytically to obtain aluminium metal. Alumina by itself also has uses in various ceramic materials. Table 2.5 presents the 2011 global alumina and aluminium production figures.

TABLE 2.5

Global Production Figures for Alumina and Aluminium (2011)

	Thousand Metric Tonnes	
Region	Alumina	Aluminium
Africa & Asia (except China)	7046	4338
North America	5723	4969
South America	15099	2185
West Europe	5849	4027
East & Central Europe	4676	4319
Oceania	19637	2306
Gulf Cooperation Council (GCC)		3483
China	34078	17786
Estimated Unreported		576
Total	92108	43989

Source: http://www.world-aluminium.org/statistics/alumina-production/#data.

The conventional process for refining bauxite ore is the Bayer process and over 80% of the world alumina production comes from this process (Tsakiridis, 2012). Alternatives, namely the Sinter process, the combined/parallel Bayer–Sinter process, and the Nepheline-based process, are also sometimes employed in order to accommodate different raw materials and to improve the recovery. The Bayer process involves digestion of the ore with sodium hydroxide. The aluminium compounds dissolve to form sodium aluminate and the remaining solid impurity (termed "red mud") is separated by settling and filtration. The sodium aluminate is crystallized, washed, and calcined to obtain alumina crystals. Aluminium production is undertaken by the Hall-Heroult process where the alumina crystals are dissolved in cryolite (sodium hexafluoroaluminate, Na_3AlF_6) and electrolyzed using carbon electrodes. Molten aluminium settles down, carbon dioxide (CO_2) is generated, and a layer of oxidized aluminium with entrapped metal aluminium and gas (termed skimming or dross) floats on the surface. In secondary aluminium production, scraps and used aluminium products are melted in a reverberatory furnace or rotary furnace under a salt layer. The salt layer is composed of NaCl and KCl, with some cryolite or CaF_2, and is required to protect the aluminium metal from oxidation. Because secondary aluminium production requires 5 to 8% of energy and around 10% of capital equipment as compared to primary aluminium production, the economics of aluminium recycling are very attractive (Plunkert, 2006).

Global aluminium consumption is growing at 4.1% annually and is estimated to reach 120 Mt by 2025 (OECD, 2010). This is nearly a threefold rise over the consumption in 2006 of 45.3 Mt. BRIC countries (Brazil, Russia, India, and China) would be the chief contributors to this increase in consumption. To meet this projected demand, approximately 570 Mt of bauxite and 230 Mt of alumina has to be produced. Thus, even with technological improvements, the

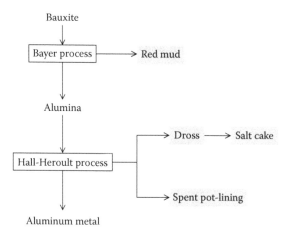

FIGURE 2.2 Solid wastes generated in aluminium production.

volume of waste generated in aluminium processing will be significant in the coming years.

Annually, over 4.5 billion kilograms of aluminium production waste is generated worldwide (Shinzato and Hypolito, 2005; Cui and Zhang, 2008). Figure 2.2 shows the main solid waste products in aluminium manufacture. The following sections describe each of these waste streams.

2.2.1 Red Mud

Red mud is the name more commonly given to bauxite residue, the by-product of bauxite refining by the Bayer process. The annual production rate of red mud is reportedly 120 million tonnes with an estimated 2.7 billion tonnes existing as a legacy of almost 120 years of alumina extraction from bauxite ore (Power et al., 2011). It is considered a hazardous material, predominantly due to its high alkalinity, and there are a number of options available for its storage and utilization (Power et al., 2011). As detailed by Power et al. (2011), historically storage in lagoons and marine disposal were commonly applied, although more recently there have been moves toward hyperbaric steam filtration and dry stacking methods which are used to reduce alkalinity and the possibility of leakage, respectively. In terms of the latter aspect, the hazardous nature of red mud storage is underlined by the infamously tragic events of Ajka in Hungary in October 2010 when approximately 1 million cubic meters of red mud waste was released from the rupture of a storage dam flooding the village of Kolontàr and resulting in a number of fatalities.

The composition of red mud is complex and is geographically and temporally variable, reflecting the composition of the bauxite from which it is derived. Frequently, composition is reported in terms of binary oxides, as illustrated in Table 2.6, although its mineralogy is more complex, as shown in Table 2.7. Tables 2.6 and 2.7 also illustrate the geographical variation of the material. In the tables, it can be seen that iron oxides/oxyhydroxides are major components of red mud and, indeed, these

TABLE 2.6

Chemical Composition of Different Bauxite Digestion Residues

	Weipa (Australia)	Trombetas (Brazil)	South Manchester (Jamaica)	Darling Range (Australia)	Iszka (Hungary)	Parnasse (Greece)
Digestion temperature	240°C	143°C	245°C	143°C	240°C	260°C
Components (wt%)						
Al_2O_3	17.2	13.0	10.7	14.9	14.4	13.0
SiO_2	15.0	12.9	3.0	42.6	12.5	12.0
Fe_2O_3	36.0	52.1	61.9	28.0	38.0	41.0
TiO_2	12.0	4.2	8.1	2.0	5.5	6.2
Na_2O	9.0	9.0	2.3	1.2	7.5	7.5
CaO	—	1.4	2.8	2.4	7.6	10.9
Others	3.5	1.0	2.8	2.4	4.9	2.3
LOI[a]	7.3	6.4	8.4	6.5	9.6	7.1

Source: Adapted from data referenced in Bánvölgyi and Huan, 2010.

[a] Loss on ignition.

components give rise to the typical red/orange coloration of the materials as well as dominating the potential utilization of this material.

Various potential applications of red mud have been investigated, which have been the subject of a recent review (Klauber et al., 2011). An analysis by area of application of 734 patents between 1964 and 2008 presented in the review revealed the following split: 33% relating to building and civil construction materials, 13% relating to application as catalyst and adsorbent, 12% relating to the production of ceramics, plastics, coatings, or pigments, 12% relating to application for waste and water treatment, 9% relating to the recovery of major metals, 7% relating to use in steel making and application as a slag additive, 4% relating to soil amendment or production, 4% relating to application as a gas scrubbing agent, 2% relating to recovery of minor elements, 2% relating to other applications, and 1% relating to application as a minor additive in various processes. Since the volume of red mud is so large, it is unlikely that any one single use would be developed for the material and, instead, its utilization is most likely to be accomplished by a range of different applications.

In terms of application, consideration must be given to both the composition and the variability of the material. For example, the radioactivity of some red muds based upon [226]Ra and [232]Th content necessitates caution in their application for building materials (Somlai et al., 2008), whereas others, which are much less radioactive, will not present the same issue. Where variability of composition is a problem, pre-treatment procedures can be undertaken to generate materials of a consistent nature. A brief summary of the literature related to selected avenues of interest for application of red mud are described next.

TABLE 2.7
Mineralogical Composition of Different Digestion Residues

Components (%)	Weipa	Trombetas	South Manchester	Darling Range	Iszka	Parnasse
Gibbsite	33.0	—	33.0	5.6	—	—
Hematite	3.5	38.0	3.5	14.5	33.0	38.0
Goethite	18.0	19.0	10.0	14.5	6.0	1.0
SAHS[a]	27.0	27.0	27.0	5.4	32.0	26.0
Illite	2.0	—	—	4.7	—	—
Boehmite	2.0	0.6	2.0	2.5	0.8	0.6
Diaspore	—	1.2	2.0	2.5	0.7	0.6
Ca-Al silicate	—	—	—	1.7	12.5	10.0
Calcium titanate	—	1.5	—	—	7.0	10.5
Calcite	0.5	1.4	—	2.3	3.0	3.6
Quartz	6.0	2.2	0.5	37.1	—	—
Anatase	2.0	2.5	6.0	1.0	—	—
Rutile	6.0	0.8	2.0	0.6	—	—
Sodium titanates	—	—	6.0	0.6	—	—
Magnetite	—	—	—	1.3	—	—
Chamosite	—	—	—	—	—	6.0
Ilmenite	—	—	—	1.0	—	—
Others	—	5.8	—	3.4	5.0	3.7

Source: Adapted from data referenced in Bánvölgyi and Huan, 2010.
[a] Sodium aluminium hydrosilicates (e.g., sodalite, cancrinite, etc.).

2.2.1.1   Civil and Building Construction and Ceramics

The application of red mud in the production of cements and concretes as well as bricks and blocks is well established and has been recently reviewed in detail where it has been shown that red mud can be used in the various different components of cements and concretes as well as in geopolymers (Klauber et al., 2011). Indeed, within that review, it has been pointed out that the incorporation of red mud into concrete via cement at a level of just 0.4% would be sufficient to consume its annual production. The iron and alumina components of red mud reportedly enhance the setting and strength properties of the cement, although the sodium content is detrimental. Red mud in conjunction with kaolin and active silica has also been studied as a suitable agent for preparing the coarse aggregates in concretes (Oliveira and Rossi, 2012). The application of red mud in the manufacture of both fired and non-fired bricks is well described (Klauber et al., 2011). In addition to its mechanical properties, it has also been reported to improve the color of brick (Kolesnikova et al., 1998).

In application to construction materials, it is important to note that leaching of metals from red mud can occur, although this is alleviated somewhat by sintering (see, for example, Ghosh et al., 2011). As mentioned previously, another aspect to be considered that is especially significant for the construction of buildings is the radioactive nature of the red mud applied.

The application of red mud to the production of ceramics is also an aspect well studied which has formed part of the comprehensive review by Klauber, Gräfe, and Power (2011). Areas of application in this context can involve being a major component of the ceramic mix or as a pigment or glazing agent. Different types of ceramic materials have been reported, for example those made entirely from red mud (Pontikes et al., 2009) and glass-ceramics from a mixture of wastes (Bernardo et al., 2009). Reduction of the radioactivity to acceptable levels in ceramics made from a radioactive red mud via the addition of barium carbonate prior to sintering has been reported recently (Qin and Wu, 2011).

2.2.1.2 Catalysis and Gas Scrubbing

The catalytic applications of red mud have been recently reviewed (Sushil and Batra, 2008). The predominant interest in terms of catalysis focuses upon the presence of iron in the material and a wide range of reactions including hydrogenation, liquefaction, hydrodechlorination, combustion, and environmental catalysis have been investigated. The material has been directly applied as an active phase itself, in modified form or as a chemically active support for other phases such as metals. When applying the material directly, account has to be taken of the geographical and temporal compositional variability of the material. This aspect is illustrated in a study of methane cracking over a series of red muds originating from India, as illustrated in Figure 2.3 (Balakrishnan et al., 2009). Within this figure, the three samples can be seen to exhibit markedly different activity profiles and it is salient to note that two of the samples originate from the same site, which was sampled at

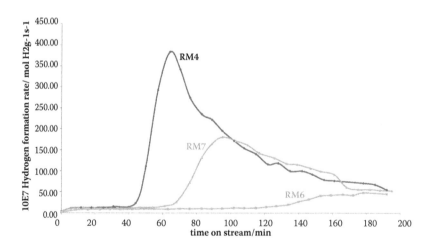

FIGURE 2.3 Comparative hydrogen formation rates from methane cracking at 800°C over three red mud samples of Indian origin (Balakrishnan et al., 2009). (Reproduced by permission of The Royal Society of Chemistry http://pubs.rsc.org/en/content/articlelanding/2009/gc/b815834g.)

different times. If the red mud were to be used in "raw" form, the likely variation in catalytic performance would need to be offset against the fact that this represents a very cheap and plentiful resource of impure, multifunctional iron oxide, although in some instances it is the basicity associated with the material which is of greater interest (e.g., Liu et al., 2012). It is possible to standardize the behavior by chemical pretreatment procedures and red mud could always, of course, be used as a resource from which the iron component is selectively extracted. Various procedures to both enhance catalytic performance and standardize the behavior of the material have been reported. For example, with respect to enhancement of performance, sulfiding red mud can lead to enhanced activity in some reactions (e.g., Ordóñez et al., 2001). A method widely applied to modify the materials for catalytic application is the dissolution of red mud in hydrochloric acid followed by reprecipitation at pH 8 with aqueous ammonia (Pratt and Christoverson, 1982), which leads to a material of higher surface area and reduced content of sodium and calcium containing components that can act as catalyst poisons. Further modification of this procedure has involved the incorporation of low quantities of phosphorus, stated to be beneficial for some reactions, through the inclusion of phosphoric acid along with hydrochloric acid in the initial dissolution step (Álvarez et al., 1999). While it may be envisaged that the application of pure phase materials are preferred for most applications, it is interesting to note that the mixture of phases present in red mud is indeed advantage for some applications. In this respect, it is interesting to note that red mud has been applied for the upgrading of pyrolysis-derived bio-oil wherein the combination of basicity and active phases achieved the hydrogenation of problematic acid components generated through fast pyrolysis (Karimi et al., 2010).

As stated previously, red mud can be viewed as a cheap and plentiful resource of impure iron oxide. It is therefore expendable. However, potential use of, or the disposal issues associated with, the post-reaction material need to be considered. When the material is exposed to a reducing atmosphere, magnetic materials such as Fe_3O_4, Fe, and Fe_3C can result and these may be of interest in further application such as, for example, magnetic ampiphilic composites leading to the separation of biodiesel from water (Oliviera et al., 2010). The cracking of hydrocarbons, waste products themselves in some circumstances, over red mud can yield magnetic composites comprising carbons of different form as illustrated in Figure 2.4 (Sushil et al., 2010). Such composites could be applied to other processes. In view of the possibility of the combination of two large-scale waste products to yield products of value, it is interesting to note that red mud is active for plastic waste pyrolysis (López et al., 2011).

The inherent alkalinity of red mud can be put to effect in the removal of acidic gases from effluent streams. In this respect, a number of studies have applied red mud to the scrubbing of SO_2. For example, there is an SO_2 scrubbing process that was developed by Sumitomo and disclosed in the patent literature (e.g., Yamada et al., 1980). As well as performing an environmentally important purpose, scrubbing also leads to a reduction in the alkalinity of the residue, which is one of the major issues associated with its handling. CO_2 scrubbing would also produce such benefits. The interaction of red mud with H_2S can also be beneficial not only in terms of effluent stream decontamination, but also in the generation of catalytically active sulfided materials as briefly mentioned before.

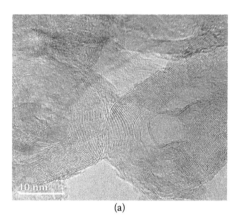

(a)

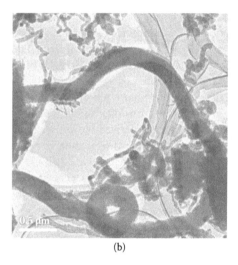

(b)

FIGURE 2.4 Morphology of carbons produced by cracking (a) methane and (b) butane over red mud (Sushil et al., 2010). (Reproduced with kind permission from Elsevier.)

2.2.1.3 Water Treatment and Use as a Soil Additive

The application of raw and modified red mud for the removal of heavy metals, anions, dyes, and organic molecules from water has been the subject of a number of reviews (e.g., Wang et al., 2008; Liu et al., 2011). Perhaps unsurprisingly, in view of the variability in the composition of red mud, a wide range of different behavior, adsorption capacities, and mechanisms for the same species in different studies can be found in the literature. Aspects of this are illustrated in Tables 2.8 and 2.9, which are adapted from the review by Wang et al. (2008).

In a number of instances, red mud has been reported to be an effective material for reducing the bioavailability of toxic ions/compounds. For example, Santona et al. (2006) have reported the immobilization of zinc, lead, and cadmium ions by binding to the cancrinite and haematite phases in red mud and acidified red

TABLE 2.8

The Adsorption Capacity of Various Red Mud/Modified Red Mud Samples for Various Metal Ions

Ion	Adsorbent	Temperature (°C)	pH	Adsorption (mg g⁻¹)
Cu(II)	Red mud—HCl	25	5–5.5	60.5–67.2
	Red mud—HCl	30	5.5	2.28
	Red mud—CaSO₄	—	5.5–6.2	19.7
Pb(II)	Red mud—HCl	25	5–5.5	173.4
	Red mud		5.5–5.9	1.88
	Red mud—HCl		5.5–5.9	0.77
	Red mud—H₂O₂ heated	30–50	4–6	66.9–71.3
Cd(II)	Red mud—HCl	25	5–5.5	106.4–108.6
	Red mud—heated	30–50	4	11.2–13.0
	Red mud	—	6	68
	Red mud	—	5.5–5.9	1.35
	Red mud—HCl	—	5.5–5.9	0.95
	Red mud—CaSO₄	—	4.7–6.2	10.6
Zn(II)	Red mud—heated	30–50	4	11.8–14.5
	Red mud	—	7	133
	Red mud	—	5.5–5.9	2.47
	Red mud—HCl	—	5.5–5.9	1.59
	Red mud—CaSO₄	—	6.9–7.8	12.6
Cr(VI)	Red mud—H₂O₂ heated	30–50	2	21.1–22.7
Ni(II)	Red mud—CaSO₄	—	7.5–7.9	10.9
As(V)	Red mud	25–70	3.5	0.51–0.82
	Red mud—HCl	25–70	3.5	0.94–1.32
	Red mud	—	6	0.9
As(III)	Red mud	25–70	7.2	0.46–0.67
	Red mud—HCl	25–70	7.2	0.34–0.88

Source: Taken from the literature as reported by Wang et al., 2008.

mud has been reported to be highly effective for the uptake of phosphate, a major cause of eutrophication (Li et al., 2006). However, there are a number of issues with the application of red mud to water treatment, which may need to be considered or overcome depending upon specific application. It is a highly alkaline material, typically resulting in pHs greater than 11, and points of zero charge reported are frequently in the region of 8.5 (Pulford et al., 2012). Figure 2.5 reproduces pH titration curves for a red mud sample, which illustrates a number of relevant issues. In Figure 2.5(a), it can be seen that the titration curves are irreversible. This results if insufficient equilibration time is given and highlights the difficulty of handling of the material in terms of its phase composition/stability. In the point of zero charge plot [Figure 2.5 (b)], the regions of buffering occurring around pH 6-8 and < pH 4 can be seen. These can be related to the presence of sodalite and

TABLE 2.9

The Adsorption Capacity of Various Red Mud/Modified Red Mud Samples for Various Dyes

Dye	Adsorbent	Temperature (°C)	Adsorption (mg g^{-1})
Congo red	Red mud	30	4.05
Acid violet	Red mud	30	1.37
Procion orange	Red mud	30	6.0
Rhodamine B	Red mud + H_2O_2	30	5.56
Methylene blue	Red mud + H_2O_2	30	16.72
Fast green	Red mud + H_2O_2	30	7.56
Methylene blue	Red mud	—	0.74
Methylene blue	Red mud	30	2.49
Methylene blue	Red mud—heat	30	0.48
Methylene blue	Red mud—HNO_3	30	1.02
Congo red	Red mud—HCl	—	7.08

Source:　Taken from the literature as reported by Wang et al., 2008.

iron-containing phases, respectively, and thus a pH dependent behavior related to mineralogical composition can be anticipated. Indeed, in a study of red mud samples originating from 11 different sources employing eight different pretreatment procedures, Snars and Gilkes (2009) have distinguished five different shapes of buffer curve. Accordingly, given such variability in composition even for the same source of sample over time, methods of pretreatment to obtain reproducible or improved performance have been explored. These have included simple washing with water (e.g., Apak et al., 1998; Cengeloglu et al., 2006), boiling or washing with acid (e.g., Apak et al., 1998; Cengeloglu et al., 2006; Santona et al., 2006), acid dissolution and reprecipitation with base (Pradhan et al., 1999), and washing with seawater (Palmer et al., 2010). A further approach adopted has been to utilize the spent magnetic composite generated by its application to hydrocarbon cracking to yield a magnetic composite of more controlled pH, which can be separated from water by application of a magnetic field (Pulford et al., 2012). Ma et al. (2012) have demonstrated the difference in copper scavenging capacity between two different red mud types, one being highly basic and the other a seawater-neutralized mud. In this study, the highly basic material was reported to have the better performance due to the presence of $CaCO_3$ and the consequent formation of atacamite (the source of Cu was $CuCl_2$).

As well as for water treatment, there has been interest in the application of red mud to soils both for remediation purposes and enhancement of alkalinity as well as enhancement of water, phosphorus, and other nutrient retention. There are a number of studies related to this aspect that have been reported in the literature. For example, Feigl et al. (2012) have reported that 5 wt% addition of red mud was effective at reducing the mobility of cadmium and zinc in contaminated soil and mine waste as well as their uptake by *Sinapis alba* test plants. A mixture of red mud and gravel

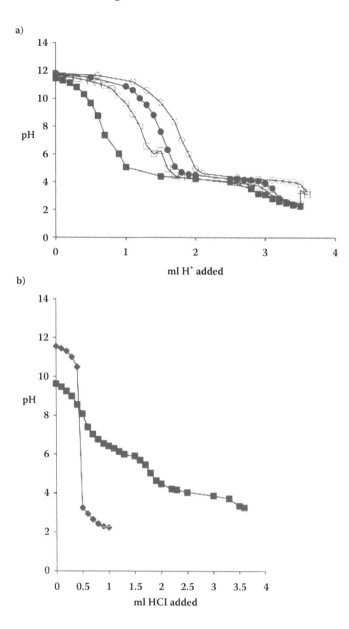

FIGURE 2.5 (a) Buffer curves for red mud titrated with 1M HCl and 1M NaOH; filled square = first addition of HCl; filled circle = first addition of NaOH; empty square = second addition of HCl; empty circle = second addition of NaOH; and (b) point of zero charge of red mud; filled squares = the titration with red mud, filled diamonds = the blank titration. The point of zero charge is given by the intersection of the two pH curves (Pulford et al., 2012). (Reproduced with kind permission from Elsevier.)

sludge has also been reported to be effective in the sulfur-enhanced phytoextraction of cadmium and zinc with *Salix smithiana*, the concept being initial immobilization followed by controlled metal solubilization (Iqbal et al., 2012). When applying red mud to soil, care needs to be taken over potential phytotoxicity as well as heavy metal leaching and radioactivity. These aspects are obviously red mud source dependent and hence there could be difficulties in translating the results obtained from one material to the behavior expected with another. In a study related to the Peel inlet and Harvey Estuary in Western Australia where red mud had been utilized to reduce algal blooms associated with phosphorus run-off from the surrounding sandy soils, Summers and Pech (1997) concluded that there was no significant increase in the discharge of heavy metals associated with the use of red mud. In a study of radionuclide uptake into food crops, again based in Western Australia and involving the application of red mud (in association with 10 wt% gypsum) for the enhancement of phosphorus retention by sandy soils, Cooper and co-workers (1995) reported that there was no significant increase in the radionuclide levels. Rates of 200 kg/ha and 800 kg/ha were applied for the lettuce and cauliflower crops, and rates of 50 and 400 kg/ha for the potato crop and cabbage growth were also investigated. The red mud applied had thorium and radium concentrations in excess of 1 kBq/kg and 300 Bq/kg, respectively. In a study relating to the effects of the Ajka red mud incident upon plant toxicity and trace metal availability in contaminated soil, Ruyters et al. (2011) have concluded that at 5 wt% and above, red mud affects the shoot yield of barley seedlings. However, the primary effect was related to the caustic soda content and although shoot copper, chromium, iron, and nickel concentrations were increased, they did not exceed toxic limits.

In addition to the various applications mentioned previously, there has also been interest in both the extraction of elements from red mud and its conversion to pig iron. Useful common metals that can be extracted from red mud are Fe, Al, Na, and Ti, possibly along with more valuable trace elements such as Nb, Ga, Zr, Th, Sc, and V (Piga et al., 1993). Various processes could be applied for this purpose, including magnetic separation to recover the iron content and the recovery of aluminium, iron, and titanium by mixing with coal, lime, and sodium carbonate and applying reducing sintering, and chlorination followed by fractional distillation or the application of a chlorine/sulfur dioxide mixture (Piga et al., 1993). Pig iron can be prepared from red mud by smelting or the use of thermal plasma technology, for example (Jayasankar et al., 2012).

2.2.2 Dross

Dross is a mixture of metallic aluminium, oxide, and salts. The dross may also contain aluminium nitrides and carbides, as well as metal oxides derived from the molten alloy. Overall, it is a heterogeneous mixture containing large lumps, fine oxides, and small pieces of metal. Due to the entrapped CO_2, the dross floats and is skimmed off. Dross is broadly categorized as "white dross" (from primary aluminium production) or "black dross" (from secondary aluminium production). Dross with high free aluminium content typically occurs as compact lumps while dross with low free metal is granular (Manfredi et al., 1997). Table 2.10 summarizes

TABLE 2.10

Comparison of Compact (White) and Granular (Black) Dross Properties

Properties	Granular Dross	Compact Dross
Alloy content (%)		
Melt	2.44–11.77	1.34–10.03
Recovered metal	1.03–5.51	0.33–6.80
Distribution(q) (mm⁻¹)	0.08 (coarse)–0.452 (fine)	—
Density (t/m³)	0.828–1.118 (bulk)	2.396–2.528 (apparent)
Metal content (%)	46.9–69.1	71–93
Lixiviate (pH)	9.52–10.14	9.03–9.48
Salt content (%)	0.18–6.21	0.01–0.03
Gas evolution (l/kg dross)	0.25–1.17	No evolution

Source: Manfredi et al., 1997.

TABLE 2.11

Elemental Composition of Dross

Element[a]	Al	Mg	Na	Si	K	Ca	Fe	Zr	Ti	F
Amount (wt%)	47.2	7.6	1	0.89	0.73	0.73	0.23	0.14	0.12	2

Source: Yoshimura et al., 2008.

[a] Others (S, Mn, Ni, Cu, Zn, P, Ga, Sr, Sn, Co, and Cr) in traces.

the physical properties of dross and Table 2.11 shows a typical elemental composition. White dross can contain 20 to 45% recoverable metallic aluminium and black dross is typically a mixture of aluminium metal (10 to 20%), aluminium oxide (20 to 50%), and a salt-flux mixture (40 to 55%) (Tsakiridis, 2012).

In general, the amount and composition of the dross generated depends upon the type of metal input (primary ore or recycled aluminium) and the process conditions (skimming, fluxing, or cooling). The amount of dross can vary from 1 to 3% of metal throughput in primary aluminium production and up to 20% in remelting operations (Lazzaro et al., 1994). It is estimated that an average of 15 to 25 kg of dross are produced per tonne of molten aluminium (Freti et al., 1982).

Because primary aluminium production is highly energy intensive and the dross contains a large fraction of metallic aluminium, metal recovery from the dross is a well-established practice in developed countries. In developing countries like India, dross treatment takes place largely in the unorganized sector (Das et al., 2007). The conventional method of dross treatment is similar to secondary aluminium production and involves heating the dross in a rotating furnace using a salt flux (usually a eutectic mixture of NaCl and KCl, with a small amount of cryolite or CaF_2). The salt flux protects the Al metal from oxidation; it also facilitates coalescence and separation of the metal from the solid oxide fraction (Tenorio and Espinosa, 2003).

TABLE 2.12

Elemental Composition of NMP

Element	Al	Mg	Na	Si	K	Ca	Fe	Cu	Pb	Zn	Cl	S
Amount (wt%)	34.4	2.44	1.69	4.4	2.31	1.32	3.6	0.99	0.14	0.6	4.23	0.07

Source: Bajare et al., 2012.

The microstructure shows that the oxide in the dross forms a continuous net entrapping the Al metal; the salt strips the oxide and promotes agglomeration of the metal (Tenorio and Espinosa, 2002). The salt-flux process is by far the most widely used as it offers the advantages of simple technology, low capital and maintenance costs, and high yield. The residue obtained after Al recovery from dross is the non-metallic products (NMP) with constituents like Al_2O_3 (corundum), $MgAl_2O_4$ (spinel), $Al(OH)_3$ (bayerite), and SiO_2 (quartz). A typical elemental composition is given in Table 2.12.

Several salt-free dross processing schemes have been developed and these have been reviewed by Unlu and Drouet (2002). The methods include the Alcan plasma torch process, the Droscar DC electric arc process, the Alurec process employing an oxygen fuel burner in a rotary furnace, the Ecocent process involving centrifuging the hot dross, and the Drosrite process where the dross is treated in a rotary furnace under an argon atmosphere. Of these options, heating by plasma arc torch (Gens, 1992; Lindsay, 1995) is promising and is in practice commercially, for example, at Alcan Treatment Center, Jonquieres, Canada.

Apart from dross treatment, total recycling of dross in electrolytic pots has also been reported on an industrial scale (Lazzaro et al., 1994). Yet another practice is to use dross for making impure chemicals and firecrackers (Das et al., 2007). In a recent report, a dross recycling facility in New Zealand is processing hot dross to recover aluminium and using the aluminium oxide residue for making fertilizer (Waste Management World, 2011).

There are continued efforts to increase the yield of metallic aluminium from dross (e.g., ALUCYC, 2011). In parallel, there is a variety of studies on developing value-added products from dross, as discussed in the following sections.

2.2.2.1 Refractory Materials and Cements

There are several patents on the use of aluminium dross for making refractory products (e.g., Yerushalmi and Sarko, 1995; Brisson et al., 1992). The use of nitride-containing aluminium dross is advantageous because the exothermic reaction between the AlN and the metal oxide (or its precursor such as silicates) cuts down the energy requirement for sintering (Brisson et al., 1992). Furthermore, the AlN can minimize contaminants like Fe_2O_3 in the metal oxide refractories by reducing them to form molten metal, which can be separated so purity and appearance can be improved.

Calcined alumina finds use in refractory applications; thus, aluminium dross has been examined as an alternative source of alumina for this purpose. Dross of particle size <700 μm can be used as a filler in concrete and asphalt to improve abrasion resistance and control microcracks (Dunster et al., 2005). However, this application is not proven. As-is or conditioned dross (obtained by washing with boiling water) used as a filler in concrete results in 25% higher flexural strength and a 5% higher compressive strength compared to pure cement (Chen, 2012). Aluminium dross from the plasma arc process contains $MgAl_2O_4$ and AlN as the main phases; this can partly replace calcined alumina as the fine structural component in castables and refractories (Yoshimura et al., 2008). Substitution up to 6.5% was studied and the physical and mechanical properties indicated that the dross could be used directly without pre-calcination. However, the mix composition has to be optimized to prevent crack-like defects in the microstructure. In fact, there are numerous studies and patents on the potential uses of NMP to produce a variety of building materials like ceramic insulation material (Garcia-Valles et al., 2008), mineral fibers with heat and sound insulating properties (Dube and Chauvette, 1991), lightweight expanded clay aggregates (Bajare et al., 2012), and refractory concrete and high alumina content cement for high-temperature applications (Shinzato and Hypolito, 2005).

As dross contains both alumina and silica, it is suitable for the synthesis of mullite ($2SiO_2 \cdot 3Al_2O_3$). Thus, mullite–zirconia composites comprising a mullite matrix with interspersed zirconia particles have been prepared from washed aluminium dross (Castro et al., 2009). Yet another material is sialon ceramics with excellent mechanical, chemical, and thermal properties. These are widely used in applications involving handling non-ferrous molten metals. β-sialon can be prepared by nitriding combustion synthesis of a mixture of aluminium dross and silicon powder (Kanehira et al., 2002). Fully dense β-sialon-AlN-polytypoid multiphase ceramics have been synthesized by hot pressing and sintering a mixture of 58% aluminium dross, 39% silicon, and 3% Si_3N_4 (Li et al., 2012).

Calcium aluminate cement mix has been prepared using aluminium sludge (containing CaO) and dross (containing Al_2O_3) (Ewais et al., 2009). Compositions in the range of 45 to 50% aluminium sludge, 37.5 to 41.25% aluminium slag, and 12.5 to 13.75% alumina were identified to be optimal as they met international standard specifications. This cement is used extensively in furnace linings for bonding refractory castables.

2.2.2.2 Chemicals and Catalysts

A process for the preparation of aluminium sulfate, a flocculant used in water clarification, by reacting the alumina constituent in the dross with sulfuric acid has been patented (Huckabay et al., 1981). Here, the concentration of contaminants has to be brought down to acceptable levels to permit utilization. By optimizing the processing conditions (concentration, time, temperature, and solids concentration), aluminium can be almost completely leached out from secondary dross by H_2SO_4 (Dash et al., 2008). Dross tailings have also been used to prepare different alums, namely aluminium-sulfate alum $[Al_2(SO_4)_3 \cdot 12H_2O]$ and ammonium-aluminium alum $[(NH_4)_2SO_4Al_2(SO_4)_3 \cdot 24H_2O]$ (Amer, 2002). The performance of such dross-derived alum for water treatment is similar to that observed with commercial alum (Mukhopadhyay et al., 2004).

Secondary aluminium dross, generated from white dross processing in a rotary DC electric arc furnace, has been subjected to alkaline leaching and carbonation (Lucheva et al., 2005). This results in aluminium hydroxide and NMP containing 80% alumina. The hydroxide is used to adjust the Al:Cl ratio in aluminium oxychloride coagulant and the NMP has applications in refractory materials.

Activated alumina has applications as an adsorbent and catalyst. In this context, η-alumina was prepared by calcining aluminium hydroxide obtained by treating dross with H_2SO_4 followed by aqueous ammonia (Das et al., 2007). Aluminate from dross tailings leached with sodium hydroxide has been precipitated by different methods, namely using H_2O_2 solution, ammonium aluminium sulfate, and active seeds of aluminium hydroxide; the precipitate was calcined at 600°C to obtain γ-Al_2O_3 (El-Katatny et al., 2003). This mesoporous γ-Al_2O_3 (average pore diameter of 62 to 69 Å) with a surface area of 158 to 245 m^2/g is a suitable catalyst support. Nanostructured alumina with crystallite size less than 100 nm has been prepared from black aluminium dross using a solvothermal method (Meor et al., 2010). Introduction of a fractional precipitation stage in the process enhanced the purity to 96.5% and produced 100% α-alumina crystalline phase.

Aluminium dross has also been used as the starting material for the preparation of crystalline microporous aluminophosphate molecular sieve $AlPO_4$-5 (Murayama et al., 2006, 2009; Kim et al., 2009). The catalytic activity appears to be unaffected by the starting material as observed with CrAPO-5 for liquid phase oxidation of tetralin using t-butyl hydroperoxide as an oxidizing agent (Kim et al., 2009). The performance of the CrAPO-5 derived from dross was the same as that with the reference catalyst made using pure chemicals. Murayama et al. (2012) have conducted extensive work on preparation of hydrotalcite-like compounds (double hydroxide composed of Mg–Al, Ca–Al, or Zn–Al) from aluminium dross. It was concluded that the presence of metal impurities in dross has little effect on the anion-exchange ability of these compounds.

In a fast, self-sustained reaction, ball milled Al dross powder was hydrolyzed to generate H_2 gas and form aluminium oxide hydroxide (AlOOH) (David and Kopac, 2012). Ball milling the dross reduces the particle size and exposes the chemically active metal surface. The hydrogen is a clean fuel and the AlOOH can be used for refractory applications or to make calcium aluminate cement.

2.2.2.3 Metal Production, Metal Casting, and Composites

Sintered calcium aluminates (that can be used as a protective cover for liquid metals, particularly steel) can be prepared from dross (Breault et al., 1995). The calcium aluminate is made by fusion of lime (CaO) and alumina (Al_2O_3) from the dross (Gens, 1992; Breault, 1995). The high content of metallic aluminium fines and aluminium oxide imparts heat generation and retention properties in NMP. This can be exploited as an exothermic topping in steel manufacture (Shinzato and Hypolito, 2005).

Aluminium dross has been used to substitute green silica sand in mould material for aluminium alloy castings (Adeosun et al., 2012a). With dross content in the 50 to 80% range, mold permeability is adversely affected. This problem can be overcome by quench tempering the cast, which results in the added benefits of improved tensile strength and microhardness of the casting.

The preparation of discontinuously reinforced metal-matrix composites (DRMMCs) from aluminium dross has been reported. Reinforcement of A356 alloy with fine dross particles (<10 μm) resulted in a small improvement in strength over the unreinforced material whereas wear resistance improved substantially with incorporation of coarse particles as received from dross (Kevorkijan, 1999). Good dispersion of the dross particles in A206 alloy has also been obtained (Chen, 2012). To enhance strength and rigidity, reinforcement of polypropylene with aluminium dross particles has been investigated (Adeosun et al., 2012b). Optimal combination of impact strength, tensile strength, and toughness was obtained with 15% loading of 150 μm dross particles or 20% addition of 53 μm particles.

2.2.2.4 Others

Dross from aluminium recycling has the texture of silt-loam soil. A mixture of dross and flue gas desulfurization product has been used as a soil substitute to remediate rock salt mounds from potash mining (Hermsmeyer et al., 2002). The salt mounds consist mainly of sodium chloride (NaCl). This cover material on the salt mounds enhances evaporation and reduces runoff; eventually, after the salt is leached off, this cover material can support vegetation.

2.2.3 SALT CAKE

To recover the aluminium entrapped in the dross, the dross is smelted under a salt cover to prevent the oxidation of the aluminium. Recovery of aluminium from scrap follows a similar smelting procedure. Salt cake (or salt slag) is the residue generated from scrap or dross smelting operations. It mainly contains sodium chloride (30 to 55%), potassium chloride (15 to 30%), aluminium oxide (15 to 30%), and residual metallic aluminium (5 to 7%) (Hwang et al., 2006; Jody et al., 1992). Other components like carbides, nitrides, sulfides, and phosphides may also be present. It is estimated that 200 to 500 kg salt cake is generated per tonne of secondary aluminium (Tsakiridis, 2012). Over 1 tonne of salt cake is produced per tonne of dross treated (Unlu and Drouet, 2002).

Salt cake is highly leachable. It is also very reactive and reacts with water or moisture in air, resulting in the formation of gases such as NH_3, CH_4, PH_3, H_2, and H_2S. On this basis, salt cake is classified as a toxic and hazardous waste as per the European Catalogue for Hazardous Wastes and the Australian Environment Protection Agency.

Because of its complex composition and the lower amounts produced (as compared to other aluminium processing wastes), there was little interest in salt cakes, and disposal in landfills was practiced in the past. This is no longer an option and may even be prohibited, as is the case in Europe. The current practice is to recover the aluminium and the salt in the slag (Hryn and Krumdick, 2002; Schwarz, 2004; Alfaro and Ballhord, 1997). Processing of salt cake or high-salt aluminium dross on a commercial scale has been reported by various companies in Europe (e.g., JBM International Ltd, UK; Engite Technologies, Italy; Alustockach, Berzelius Umwelt-Service AG (B.U.S.), Kali & Salz AG, and Alsa Technologies, all in Germany; KVS-Ekodivize, Czech Republic; RVA, France; and Befesa Escorias Salinas SA, Spain) as well as in other

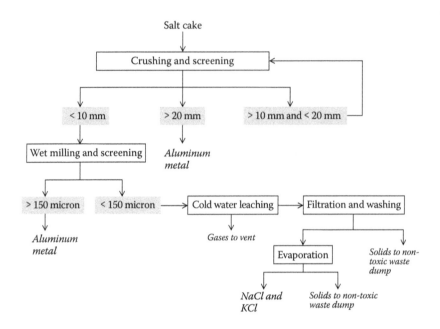

FIGURE 2.6 Treatment of salt cake. (Adapted from Bruckard and Woodcock, 2009.)

regions (e.g., Alreco Pty Ltd, Australia and Alumitech Inc. (Aleris), United States) (Bruckard and Woodcock, 2009; Tsakiridis, 2012).

The conventional salt cake treatment involving recovery of the salt and metal is shown in Figure 2.6. The recovered aluminium metal and the salts are recycled for smelting in rotary furnaces. The non-metallic residue (NMR) has high alumina content; it also has high chloride content. Washing the NMR can bring down the chloride content to acceptable level. NMR can be utilized as a source of alumina in various applications, namely clinker burning in cement kilns, preparation of ceramics, refractories and insulation (e.g., building bricks, firebricks, and mineral wool), production of chemicals (e.g., hydrate aluminium oxide and aluminium salts), and civil works (e.g., as inert fillings) (Tsakiridis, 2012).

2.2.4 Spent Pot-Lining (SPL)

The spent carbon electrodes and refractory lining from the electrolysis cells (or pots) in the Hall-Heroult process constitutes the spent pot-lining (SPL). The typical composition of this graphite-ceramics waste is given in Table 2.13. Approximately 25 kg SPL is generated per tonne of aluminium produced; this translates into a global production of nearly 1 million tonnes of SPL (Pawlek, 2012). Because it contains water soluble fluorides (up to 13.5 wt%) and cyanides (up to 1 wt%) (Chanania and Eby, 2000), SPL is categorized as a hazardous waste in many countries.

In the past, landfilling was the main method of managing SPL waste and controlled storage in secure landfills continues to be practiced (Lisbona et al., 2012). However, the fluorides and cyanides have to be removed and the waste rendered

TABLE 2.13

Composition of SPL

Component	Both Cuts Together	1st Cut Carbonaceous	2nd Cut Refractory
	100%	56%	44%
Carbon	33.1	54–64	18.2
Total fluorides	15.7	6–20	4–10
Free alumina	22.3	0–15	10–50
Total aluminum	15.1	5–15	12.6
Total sodium	14.2	5–12	12.0
Calcium	1.8	0.5–4	1–6
Quartz	2.7	0–6	10–50
Phosphorus	0.3	0–650 g/t	0–300 g/t
Sulfur	0.1	0.1	0.1

Source: von Krüger, 2011.

inert prior to landfilling. Inertization typically involves heating to 750 to 1000°C (to remove volatiles) and treatment with chemicals like lime and $FeSO_4$ (to bind fluorides/neutralize cyanide) (Pawlek, 2012).

Options for treatment and disposal of SPL have been reviewed and updated over the last two decades (Øye, 1994; Pawlek, 2012). Current trends encourage schemes that recover resources or result in value-added products from SPL. It is estimated that Asia Pacific Partnership (APP) comprising Australia, Canada, China, India, Japan, Korea, and the United States recycled 53% of the SPL produced in 2011 (International Aluminium Institute, 2011). The Ausmelt pyrometallurgical process produces aluminium fluoride (that is reused in the smelting process) and synthetic sand (that can be used for road-making and concrete production) from SPL (Matusewicz and Roberts, 2002). Alcoa has used this scheme since 2001 to process 12,000 tonnes of SPL per year. The Rio Tinto Alcan hydrometallurgical low caustic leaching and liming (LCLL) process recovers a carbonaceous by-product (used as fuel in cement plants), caustic liquor containing 27% NaOH (that can be reused in alumina refining), and CaF_2 (used in high precision optical equipment). An 80,000 tonnes per year capacity LCLL plant at Saguenay (Quebec, Canada) has been operational since 2008 (Rio Tinto, 2011). Alouette Aluminium Smelter in Canada uses the refractory material in SPL for making concrete blocks (Pawlek, 2012). In Brazil, SPL has been used in cement kilns with the first cut (carbonaceous component) acting as fuel and the second cut (refractory component) serving as clinker aid (Venancio et al., 2010).

2.2.4.1 Developments in SPL Utilization

The carbon-rich "first cut" is a good fuel that can be co-combusted with coal in thermal power plants and the resulting ash stabilized with gypsum for landfilling (Miksa et al., 2003). The carbon can also be recovered and used as an adsorbent (Mazumder and Devi, 2008). Heating the SPL with silica results in silicon carbide, which has multiple uses as a refractory, abrasive, and in power electronics (Brosnan, 2002).

The refractory-rich second cut in the SPL has been recommended for the production of Al-Si alloys (Moxnes et al., 2003), foam-silicate for thermal insulation (Proshkin et al., 2002), and red brick for construction (Miksa et al., 2003). The mobility of the cyanide and fluoride constituents can be reduced in a concrete matrix; thus, SPL mixed concrete blocks are a potential product as well (Silveira et al., 2003).

Yet another application of SPL is in pig iron manufacture where the addition of 5 to 25 kg SPL per tonne of pig iron serves both as fuel and as an agent for reducing slag viscosity (Nikitin et al., 2001). Simultaneous detoxification and utilization of SPL can be achieved by treating a mixture of SPL and iron and steel industry waste (e.g., mill scale, blast furnace sludge, etc.) in a conventional electric arc furnace (O'Connor et al., 2002). Iron, slag free of hazardous constituents, and NaF can be recovered from this process. Using plasma treatment, SPL can be directly converted to a disposable vitrified product (Tetronics, 2011).

SPL can also be co-treated with wastewater and caustic, both obtained from aluminium anodizing industry, to recover AlF_3 that can be reused in primary smelting (Lisbona et al., 2012). The scheme involves (1) leaching the fluoride with wastewater, (2) AlF_2OH precipitation with caustic, and (3) conversion of AlF_2OH to AlF_3. Acid leaching of SPL followed by hydrothermal precipitation also results in AlF_3 (Jenkins, 1994). Recovery of cryolite (96% purity) and carbon (72% purity) from SPL can be achieved using a sequential alkaline-acidic leaching process (Shi et al., 2012).

2.3 COPPER

Copper ore is typically a sulfide ore and is first enriched by froth floatation. The enriched ore is subjected to smelting where a copper-rich matte is obtained, which also contains iron sulfide. During this stage, the remaining iron in the ore is oxidized and reacts with silica flux to form a slag. The copper matte is taken to a converter where oxidation of iron takes place and 99% copper is obtained. In this stage as well, slag is generated by the reaction between iron oxide and silica flux. The slag at this stage also contains copper, which is removed by settling or froth floatation. The blister copper obtained from the converter is first refined to remove S and O and cast to form anodes. The anode is then used in electrorefining to obtain copper of high purity with less than 20 ppm impurities. Figure 2.7 shows the schematic for

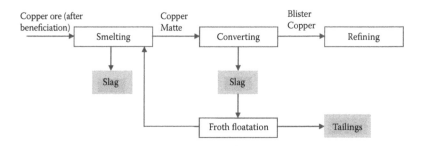

FIGURE 2.7 Copper-making process showing the wastes generated.

TABLE 2.14
Copper Production in 2011 (in thousand metric tonnes)

Region	Refined Copper Production	Refined Copper Usage
Africa	960.7	281.5
America	5423.4	2802.3
Asia	8991.8	11007.0
Europe	3796.0	4497.2
Oceania	476.8	119.9
World	19648.7	19864.5

Source: International Copper Study Group, 2012.

copper production from ore. The amount of copper production and consumption is shown in Table 2.14.

2.3.1 SLAG

Slag is the main waste that is generated during copper production from smelter and converter. Production of 1 tonne of copper generates approximately 2 to 3 tonnes of slag (Gupta et al., 2012) with approximately 24.6 million tonnes being generated globally each year (Biswas and Satapathy, 2010). The utilization of copper slag has covered many applications and these have been reviewed (Gorai et al., 2003; Wang et al., 2011).

The slag from the convertor process for copper has been used to recover the remnant copper using ferric leaching in which a biooxidation step has been incorporated to regenerate the iron to ferric form for reuse. The extent of recovery was dependent on the extraction conditions and more than 93% recovery could be obtained in 4 h (Carranza et al., 2009). Since Co is present with copper ore, the recovery of cobalt from converter slag by reduction smelting has been examined. The slag along with activated carbon as the reducing agent and CaO or TiO_2 as slag modifier was heated to different temperatures. The cobalt was recovered in the form of an alloy with iron and copper along with metallic copper. Under optimum conditions, approximately 94% cobalt could be recovered (Zhai et al., 2011).

Reduction smelting to recover iron from copper slag has also been studied. It was observed that by increasing the basicity with the addition of calcium oxide, the reduction temperature could be brought down. Under optimum conditions, approximately 92% iron could be recovered (Lei et al., 2010). Reduction smelting under nitrogen atmosphere in laboratory tests revealed the formation of iron copper alloy with precious metals, slag with low metal content, and gas with carbon monoxide, which could be used as fuel (Gonzalez et al., 2005). Oxidative treatment of slag from copper matte smelting has been studied. The target was to separate the iron and silica components, which were present as faylite in the slag ($2FeO \cdot SiO_2$), to hematite and amorphous silica. These could then be further processed to recover the ferrous, nonferrous, and silicate phases (Gyurov et al., 2011).

The extraction of non-ferrous metals from copper-nickel converter slag using sulfuric acid has been examined. The metals can then be further separated by liquid-liquid extraction. It was observed that the rate and extent of extraction of metal was more at a higher temperature of 400°C (Kopkova et al., 2011). Acid roasting of converter slag followed by hot water leaching was used to recover copper, cobalt, and zinc. It was found that the roasting temperature influenced the recovery of the metals, as did thermal decomposition treatment before leaching (Arslan and Arslan, 2002). Oxidative leaching using sulfuric acid and hydrogen peroxide was used to recover non-ferrous metals from copper smelter slag without the formation of silica gel, which could hinder the pulp filtration. This was followed by selective solvent extraction and 80% copper, 90% cobalt, and 90% zinc could be recovered (Banza et al., 2002). In order to selectively leach base metals such as copper, cobalt, and zinc from copper smelter slag, additives have been added to sulfuric acid. The additives include sodium chlorate to oxidize the ferrous sulfate formed and calcium hydroxide to neutralize and precipitate the silica in the slurry. By this procedure approximately 98% cobalt, 97% zinc, and 89% copper could be recovered; the dissolution of iron and silicon was only 0.02% and 3.2%, respectively (Yang et al., 2010).

Many studies have examined the use of copper slag in cement and concrete and it has been reviewed recently (Shi et al., 2008). Copper slag has been used as a replacement for fine aggregate in mortar and concrete. It was found that replacement in the range of 40 to 50% could be done with improvement in mechanical properties (Al-Jabri et al., 2011). Copper slag due to its pozzolanic properties has been used as an additive in blended cements (de Rojas et al., 2008). Copper slag has been studied as a partial replacement of Portland cement in concrete mixtures; it was found that in amounts up to 5%, there was no deterioration in properties. At higher amounts, there was deterioration of strength (Al-Jabri et al., 2005). Geopolymerization binds heavy metals better than Portland cement in toxic wastes. Thus, geopolymerization of Cu-Ni slag was studied. It was observed that the mechanical activation of the slag prior to geopolymer preparation led to better mechanical strength. The increase was more when the activation was done in a CO_2 atmosphere (Kalinkin et al., 2012).

Copper slag has been used to stabilize expansive clay soils that cause problems for road and other construction applications (Gupta et al., 2012). Copper slag has also been found to be suitable as a fill material for land reclamation and the leaching of heavy metals was within limits (Lim and Chu, 2006).

The addition of boron-containing colemanite ($2CaO_3 B_2O_3 5H_2O$) to reduce copper content in slag during copper matte smelting has been studied. Different amounts of colemanite were added to a mixture of synthetic slag and matte and comelted. It was observed that the addition could reduce the copper content in slag from 1.5% to 0.4% (Rusen et al., 2012). Copper slag has been used as filler in epoxy glass composites. The tensile strength and flexural strength increased with the addition of copper slag up to an optimum amount beyond which it came down (Biswas and Satapathy, 2010). The iron content in the slag has led to the use of slag as a catalyst for wet oxidation of phenol in the presence of hydrogen peroxide and UV light. It was found that although toxic compounds were formed, they could be minimized by controlling the reaction time (Huanosta-Gutierrez et al., 2012).

2.3.2 FLOATATION WASTE

Sulfuric acid leaching in the presence of ferric ions was carried out on floatation tailings of copper converter slag to recover copper and zinc, where the ferric iron was generated by biooxidation of ferrous iron (Muravyov et al., 2012). Fly ash has been mixed with floatation waste from copper smelting to reduce the leaching of zinc, lead, and cobalt. It was observed that the addition of fly ash reduced the extent of heavy metal leaching because of metal ion adsorption on fly ash (Coruh, 2012). The waste from floatation of copper slag was used as raw material for Portland cement clinker. There was no deterioration in properties and the leaching of heavy metals was within limits (Alp et al., 2008). Floatation waste has been vitrified with other additives to obtain glass and glass ceramics. The leaching from these was within limits and the properties showed potential for application in construction and the building sector (Coruh et al., 2006).

2.4 NICKEL

Nickel is commonly extracted from sulfide ore and the procedures are very similar to copper extraction in terms of smelting, converting, and refining. Nickel is also present in the oxide form in laterite ore from which nickel is extracted in the form of ferro-nickel by smelting. Slag is therefore a primary waste from nickel production.

Metal recovery from slag has been examined in some studies. The slag has been subjected to leaching using an aqueous solution containing $2\,wt\%$ SO_2. Approximately 80% cobalt, 60% iron, 35% nickel, and 68% zinc could be extracted. Even though copper dissolution was rapid initially, its precipitation took place (Ahmed et al., 2010). Oxidative leaching of similar smelter slag with sulfuric acid under pressure has been reported. Under optimum conditions, very high recoveries (>90%) were obtained for zinc, copper, nickel, and cobalt. In addition, the iron content in the liquid was low and the sulfidic sulfur content of the residue was negligible (Li et al., 2008). Under similar leaching procedures, the characteristics of the residue were analyzed. At low acid additions, the residue contained crystalline hematite in amorphous silica, which showed good filterability (Li et al., 2009). Nickel slag has been used along with BF slag and sand to prepare glass ceramics. The crystalline phase present was augite (Wang et al., 2011). Slag from ferro-nickel production has been used along with fly ash and steel slag for making glass ceramics by the vitrification procedure. The leaching from the produced material was low and it had good hardness (Karamberi and Moutsatsou, 2012).

2.5 OTHER METAL WASTES

The other metal wastes that have been examined are from production of lead and zinc. Lead extraction follows reduction in a blast furnace similar to pig iron making. The resulting slag has been studied as a replacement for sand in cement mortar. It was observed that up to 35% slag could be used and the leaching of heavy metals was within limits (Saikia et al., 2012). In another recent study, lead slag obtained from a laboratory process involving melting of oxidized lead concentrate was studied.

The slag along with fly ash was used to prepare geopolymers with acceptable mechanical properties (Onisei et al., 2012). The recovery of metals such as germanium, zinc, and arsenic from BF slag has been studied. The slag was first smelted under reducing conditions to obtain an alloy with iron and these metals. The alloy was subjected to two-step alkaline and acid leaching to recover the germanium and arsenic. Zinc was released as vapors during the smelting step and recovered (Cengizler and Eric, 2011).

The common procedure for zinc extraction is leaching of ore with sulfuric acid and extraction of metal by electrowinning. In this process, the solution is refined to remove impurities that are precipitated by cementation. This leads to formation of a waste solid high in zinc, cadmium, and copper. Separation of cadmium and zinc from this waste was studied by leaching with the zinc-depleted solution from electrolysis stage. Under optimum cementation conditions, the cadmium was obtained as a precipitate and the zinc-rich solution was suitable for use in the electrolysis step (Gouvea and Morais, 2007). Zinc is also produced using the imperial smelting route where the reaction of the ore with reducing CO gas forms metallic zinc vapors from which zinc and molten lead are collected. The slag from this process has been examined as an aggregate in concrete. The results were promising and no expansion due to alkali silica reactions was observed (Morrison and Richardson, 2004).

REFERENCES

Adeosun, S.O., Sekunowo, O.I., Balogun, S.A., and Obembe, O.O. 2012a. Study on the mechanical properties of cast 6063 Al alloy using a mixture of aluminum dross and green sand as mold. *JOM* 64: 905–910.

Adeosun, S.O., Usman, M.A., Ayoola, W.A., and Sekunowo, I.O. 2012b. Evaluation of the mechanical properties of polypropylene-aluminum-dross composite. *ISRN Polymer Science*, Article ID 282515, p. 6.

Agrawal, R.K. and Pandey, P.K. 2005. Productive recycling of basic oxygen furnace sludge in integrated steel plant. *Journal of Scientific and Industrial Research* 64: 702–706.

Ahmed, I.B., Gbor, P.K., and Jia, C.Q. 2010. Aqueous sulphur dioxide leaching of Cu, Ni, Co, Zn and Fe from smelter slag in absence of oxygen. *The Canadian Journal of Chemical Engineering* 78: 694–703.

Alanyali, H., Col, M., Yılmaz, M., and Karagoz, S. 2006. Application of magnetic separation to steelmaking slags for reclamation. *Waste Management* 26: 1133–1139.

Alfaro, I. and Ballhord, R. 1997. The applications of aluminum-oxide obtained from the recycling of aluminium. In: *Proceedings of the Third International Conference on Recycling of Metals*, June 11–13, 1997, Barcelona, Spain, pp. 405–441.

Al-Jabri, K.S., Al-Saidy, A.H., and Taha, R. 2011. Effect of copper slag as a fine aggregate on the properties of cement mortars and concrete. *Construction and Building Materials* 25: 933–938.

Al-Jabri, K.S., Taha, R.A., Al-Hashmi, A., and Al-Harthy, A.S. 2005. Effect of copper slag and cement by-pass dust addition on mechanical properties of concrete. *Construction and Building Materials* 20: 322–331.

Alp, I., Deveci, H., and Sungun, H. 2008. Utilization of flotation wastes of copper slag as raw material in cement production. *Journal of Hazardous Materials* 159: 390–395.

ALUCYC. 2011. Development of new technology for aluminium dross complete recovery, http://cordis.europa.eu/search/index.cfm?fuseaction = proj.document&PJ_RCN = 12396801.

Álvarez, J., Ordóñez, S., Rosal, R., Sastre, H., and Díez, F.V. 1999. A new method for enhancing the performance of red mud as a hydrogenation catalyst. *Applied Catalysis A: General* 180: 399–409.

Amer, A.M. 2002. Extracting aluminum from dross tailings. *JOM* 54: 72–75.

Apak, R., Guclu, K., and Turgut, M.H. 1998. Modelling of copper (II), cadmium (II) and lead (II) adsorption on red mud. *Journal of Colloid and Interface Science* 203: 122–130.

Arslan, C. and Arslan, F. 2002. Recovery of copper, cobalt, and zinc from copper smelter and converter slags. *Hydrometallurgy* 67: 1–7.

Australasia Slag Association. 2010. http://www.asa-inc.org.au/documents/ASA_survey_results_2010.pdf.

Bajare, D., Korjakins, A., Kazjonovs, J., and Rozenstrauha, I. 2012. Pore structure of lightweight clay aggregate incorporate with non-metallic products coming from aluminium scrap recycling industry. *Journal of the European Ceramic Society* 32: 141–148.

Balakrishnan, M., Batra, V.S., Hargreaves, J.S.J., Monaghan, A., Pulford, I.D., Rico, J.L., and Sushil, S. 2009. Hydrogen production from methane in the presence of red mud–making mud magnetic. *Green Chemistry* 11: 42–47.

Bánvölgyi, G. and Huan, T.M. 2010. De-watering, disposal and utilization of red mud: state of the art and emerging technologies, proceedings of xviii international symposium of ICSOBA 25th –27th November 2010, Zhengahoo, China, 431–443.

Banza, A.N., Gock, E., and Kongolo, K. 2002. Base metals recovery from copper smelter slag by oxidizing leaching and solvent extraction. *Hydrometallurgy* 67: 63–69.

Baricová, D., Pribulová, A., Demeter, P., Buko, B., and Rosova, A. 2012. Utilizing of the metallurgical slag for production of cementless concrete mixtures. *Metalurgija* 51(4): 465–468.

Barišić, I., Dimter, S., and Netinger, I. 2010. Possibilities of application of slag in road construction. *Technical Gazette* 17: 523–528.

Bernardo, E., Esposito, L., Rambaldi, E., Tucci, A., Pontikes, Y., and Angelopoulos, G.N. 2009. Sintered esseneite-wollastonite-plagioclase glass-ceramics from vitrified waste. *Journal of the European Ceramic Society* 29: 2921–2927.

Bhattacharya, A.K., Naiya, T.K., Mandal, S.N., and Das, S.K. 2008. Adsorption, kinetics and equilibrium studies on removal of Cr(VI) from aqueous solutions using different low-cost adsorbents. *Chemical Engineering Journal* 137: 529–541.

Bialuch, R., Merkel, T., and Motz, H. 2011. European environmental policy and its influence on the use of slag products. Second International Slag Valorisation Symposium, The Transition to Sustainable Materials Management, April 18–20, 2011, Leuven, Belgium.

Biswas, S. and Satapathy, A. 2010. Use of copper slag in glass-epoxy composites for improved wear resistance. *Waste Management & Research* 28: 615–625.

Boehme, L. and Hende, D.V.D. 2011. Ferromolybdenum slag as valuable resource material for the production of concrete blocks. Second International Slag Valorisation Symposium, The Transition to Sustainable Materials Management, April 18-20, 2011, Leuven, Belgium, pp. 129–143.

Bowden, L.I., Jarvis, A., Younger, J.P., and Johnson, K L. 2009. Phosphorus removal from waste waters using basic oxygen steel slag. *Environmental Science & Technology* 43: 2476–2481.

Breault, R., Tremblay, S.P., Huard, Y., and Mathieu, G. 1995. Process for the preparation of calcium aluminates from aluminium dross residues. US Patent 5,407,459.

Brisson, C., Chauvette, G., Kimmerle, F.M., and Roussel, R. 1992. Process for using dross residues to produce refractory products. US Patent 5,132,246.

Brosnan, D.A. 2002. Process for recycling spent pot liner. US patent 6471931.

Bruckard, W.J. and Woodcock, J.T. 2009. Recovery of valuable materials from aluminium salt cakes. *International Journal of Mineral Processing* 93: 1–5.

Camci, L., Aydin, S., and Arslan, C. 2002. Reduction of iron oxides in solid wastes generated by steel works. *Turkish Journal of Engineering and Environmental Sciences* 26: 37–44.

Carranza, F., Romero, R., Mazuelos, A., Iglesias, N., and Forcat, O. 2009. Biorecovery of copper from converter slags: slags characterization and exploratory ferric leaching tests. *Hydrometallurgy* 97: 39–45.

Castro, M.N.I., Robles, A.J.M., Hernández, C.D.A., Bocardo, E.J.C., and Torres, T.J. 2009. Development of mullite/zirconia composites from a mixture of aluminum dross and zircon. *Ceramics International* 35: 921–924.

Cengeloglu, Y., Tor, A., Ersoz, M., and Turgut, M.H. 2006. Removal of nitrate from aqueous solution by using red mud. *Separation and Purification Technology* 51: 374–378.

Cengizler, H. and Eric, R.H. 2011. Recovery of germanium from lead blast furnace slag. Second International Slag Valorisation Symposium, The Transition to Sustainable Materials Management, April 18–20, 2011, Leuven, Belgium.

Chanania, F. and Eby, E. 2000. Proposed best demonstrated available technology (BDAT) background document for spent aluminum potliners. K088 Office of Solid Waste; US Environmental Protection Agency, Washington, DC.

Chen, D., 2012. Development of aluminum dross-based material for engineering applications, Master's Thesis, Worcester Polytechnic Institute.

Cheng, T.W. and Chiu, J.P. 2003. Fire-resistant geopolymer produced by granulated blast furnace slag. *Minerals Engineering* 16: 205–210.

Cooper, M.B., Clarke, P.C., Robertson, W., McPharlin, I.R., and Jeffrey, R.C. 1995. An investigation of radionuclide uptake into food crops grown in soils treated with bauxite mining residues. *Journal of Radioanalytical and Nuclear Chemistry* 194: 379–387.

Coruh, S. 2012. Leaching behavior and immobilization of copper flotation waste using fly ash. *Environmental Progress & Sustainable Energy* 31(2): 269–276.

Coruh, S., Ergun, O.N., and Cheng, T.W. 2006. Treatment of copper industry waste and production of sintered glass–ceramic. *Waste Management & Research* 24: 234–241.

Cui, J. and Zhang, L. 2008. Metallurgical recovery of metals from electronic waste: a review. *Journal of Hazardous Materials* 158: 228–256

Das, B., Prakash, S., Reddy, P.S.R., and Misra, V.N. 2007. An overview of utilization of slag and sludge from steel industries. *Resources, Conservation and Recycling* 50: 40–57.

Das, B.R., Dash, B., Tripathy, B.C., Bhattacharya, I.N., and Das, S.C. 2007. Production of η-alumina from waste aluminium dross. *Minerals Engineering* 20: 252–258.

Das, S.K., Kumar, S., and Ramachandrarao, P. 2000. Exploitation of iron ore tailing for the development of ceramic tiles. *Waste Management* 20(8): 725–729.

Dash, B., Das, B.R., Tripathy, B.C., Bhattacharya, I.N., and Das, S.C. 2008. Acid dissolution of alumina from waste aluminium dross. *Hydrometallurgy* 92: 48–53.

David, E. and Kopac, J. 2012. Hydrolysis of aluminum dross material to achieve zero hazardous waste. *Journal of Hazardous Materials* 209–210: 501–509.

Davidovits, J. 2011. Application of Ca-based geopolymer with blast furnace slag, a review. Second International Slag Valorisation Symposium, The Transition to Sustainable Materials Management, April 18–20, 2011, Leuven, Belgium.

de Rojas, M.I.S., Rivera, J., Frias, M., and Marin, F. 2008. Review use of recycled copper slag for blended cements. *Journal of Chemical Technology and Biotechnology* 83: 209–217.

Dippenaar, R. 2004. Industrial uses of slag—the use and re-use of iron and steelmaking slags. VII International Conference on Molten Slags Fluxes and Salts, The South African Institute of Mining and Metallurgy, pp. 57–70.

Dube, G. and Chauvette, G. 1991. Process for producing mineral fibers incorporating an alumina-containing residue from a metal melting operation and fibers so produced. US Patent 5,045,506.

Dunster, A.M., Moulinier, F., Abbott, B., Conroy, A., Adams, K., and Widyatmoko, D. 2005. Added value of using new industrial waste streams as secondary aggregates in both concrete and asphalt. The Waste & Resources Action Programme, Banbury, Oxon, UK. http://www2.wrap.org.uk/downloads/BRE_Added_value_study_report.5f7f6075.1753.pdf.

Durinck, D., Engstrom, F., Arnout, S., Heulens, J., Jones, P.T., Bjorkman, B., Blanpain, B., and Wollants, P. 2008. Hot stage processing of metallurgical slags. *Resources, Conservation and Recycling* 52: 1121–1131.

Dziarmagowski, M. 2005. Possibilities of converter slag utilization. *Archives of Metallurgy and Materials* 50(3): 769–782.

El-Katatny, E.A., Halawy, S.A., Mohammed, M.A., and Zaki, M.I. 2003. Surface composition, charge and texture of active alumina powders recovered from aluminum dross tailings chemical waste. *Powder Technology* 132: 137–144.

El-Mahllawy, M.S. 2008. Characteristics of acid resisting bricks made from quarry residues and waste steel slag. *Construction and Building Materials* 22: 1887–1896.

Engström, F., Pontikes, Y., Geysen, D., Jones, P.T., Björkman, B., and Blanpain, B. 2011. Review: hot stage engineering to improve slag valorisation options. Second International Slag Valorisation Symposium, The Transition to Sustainable Materials Management, April 18–20, 2011, Leuven, Belgium.

Ewais, E.M.M., Khalil, N.M., Amin, M.S., Ahmed, Y.M.Z., and Barakat, M.A. 2009. Utilization of aluminum sludge and aluminum slag (dross) for the manufacture of calcium aluminate cement. *Ceram. Int.* 35: 3381–3388.

Feigl, V., Anton, A., Uzinger, N., and Gruiz, K. 2012. Red mud as a chemical stabilizer for soil contaminated with toxic metals. *Water Air Soil Pollution* 223: 1237–1247.

Fredericci, C., Zanotto, E.D., and Ziemath, E.C. 2000. Crystallization mechanism and properties of a blast furnace slag glass. *Journal of Non-Crystalline Solids* 273: 64–75.

Freti, S., Bornand, J.D., and Buxman, K. 1982. Metallurgy of dross formation on aluminium melts. *Light Metals* 100: 3–16.

Gao, J., Li, S., Zhang, Y., Zhang, Y., Chen, P., and Shen, P. 2011. Process of re-resourcing of converter slag. *Journal of Iron and Steel Research International* 18(12): 32–39.

Garcia-Valles, M., Avials, G., Martinez, S., Terradas, R., and Nogues, J.M. 2008. Acoustic barriers obtained from industrial waste. *Chemosphere* 72: 1098–1102.

Gens, T. 1992. Recovery of aluminum from dross using the plasma torch. US Patent 5,135,565.

Ghosh, I., Guha, S., Balasubramaniam, R., and Ramesh Kumar, A.V. 2011. Leaching of metals from fresh and sintered red mud. *Journal of Hazardous Materials* 185: 662–668.

Gomez, G.S.L. and Santoz, V.L.D. 2012. Systems and methods for recycling steelmaking converter sludge. US Patent 0167714A1.

Gonzalez, C., Parra R., Klenovcanova, A., Imris, I., and Sanchez, M. 2005. Reduction of Chilean copper slags: a case of waste management project. *Scandinavian Journal of Metallurgy* 34: 143–149.

Gorai, B., Jana, R.K., and Premchand. 2003. Characteristics and utilisation of copper slag a review. *Resources, Conservation and Recycling* 39: 299–313.

Gouvea, L.R. and Morais, C.A. 2007. Recovery of zinc and cadmium from industrial waste by leaching/cementation. *Minerals Engineering* 20: 956–958.

Gupta, R.C., Thomas, B.S., Gupta, P., Rajan, L., and Thagriya, D. 2012. An experimental study of clayey soil stabilized by copper slag. *International Journal of Structural & Civil Engineering Research* 1(1): 110–119.

Gyurov, S., Kostova, Y., Klitcheva, G., and Ilinkina, A. 2011. Thermal decomposition of pyrometallurgical copper slag by oxidation in synthetic air. *Waste Management & Research* 29(2): 157–164.

Hermsmeyer, D., Diekmann, R., van der Ploeg, R.R., and Horton, R. 2002. Physical properties of a soil substitute derived from an aluminum recycling by-product. *Journal of Hazardous Materials B* 95: 107–124.

Hisyamudin, M.N.H., Yokoyama, S., and Umemoto, M. 2009. Utilization of EAF reducing slag from stainless steelmaking process as a sorbent for CO_2. *World Academy of Science, Engineering and Technology* 32: 429–434.

Hryn, J.N. and Krumdick, G.K. 2002. Recycling aluminum salt cake, in light metals. Proceedings of Sessions, TMS Annual Meeting, Warrendale, PA.

Huanosta-Gutierrez, T., Dantas, R.F., Ramirez-Zamora, R.M., and Esplugas, S. 2012. Evaluation of copper slag to catalyze advanced oxidation processes for the removal of phenol in water. *Journal of Hazardous Materials* 213–214: 325–330.

Huckabay, J.A., Durward, A., and Skiathas, A.D. 1981. Aluminium dross processing. US Patent 4,252,776.

Hwang, J.Y., Huang, X., and Xu, Z. 2006. Recovery of metals from aluminium dross and salt cake. *Journal of Minerals and Materials Characterization and Engineering* 5: 47–62.

Iacobescu, R.I., Koumpouri, D., Pontikes, Y., Şaban, R., and Angelopoulos, G. 2011. Utilization of EAF metallurgical slag in "green" belite cement. *U.P.B. Science Bulletin Series B* 73: 187–194.

International Aluminium Institute. http://www.world-aluminium.org/statistics/alumina-production/#data.

International Aluminium Institute. 2011. Report on Aluminium Measuring & Benchmarking 2010. http://www.world-aluminium.org/media/filer_public/2013/01/15/final_app_benchmarking_and_measurement_project_2010_data_10-11-2011.pdf.

International Copper Study Group. 2012. http://www.icsg.org/index.php/component/jdownloads/finish/165/872.

Iqbal, M., Puschenreiter, M., Oburger, E., Santer, J., and Wenzel, W.W. 2012. Sulfur-aided phytoextraction of Cd and Zn by Salix smithiana combined with in situ metal immobilization by gravel sludge and red mud. *Environmental Pollution* 170: 221–231.

Jayasankar, K., Ray, P.K., Chaubey, A.K., Padhi, A., Satapathy, B.K., and Mukherjee, P.S. 2012. Production of pig iron from red mud waste fines using thermal plasma technology. *International Journal of Minerals, Metallurgy and Materials* 19: 679–684.

Jenkins, D.H. 1994. Recovery of aluminium and fluoride values from spent pot lining. US Patent 5352419.

Jody, B.J., Daniels, E.J., Bonsignore, P.V., and Karvelas, D.E. 1992. Recycling of aluminum salt cake. *Journal of Resource Management Technology* 20: 38–49.

Kalinkin, A.M., Kumar, S., Gurevich, B.I., Alex, T.C., Kalinkina, E.V., Tyukavkina, V.V., Kalinnikov, V.T., and Kumar, R. 2012. Geopolymerization behavior of Cu–Ni slag mechanically activated in air and in CO_2 atmosphere. *International Journal of Mineral Processing* 112–113: 101–106.

Kalyoncu, R.S. 2000. Slag—iron and steel. *U.S. Geological Survey Minerals Yearbook.*

Kanehira, S., Miyamoto, Y., Hirota, K., and Yamaguchi, O. 2002. Recycling of aluminum dross to sialon-based ceramics by nitriding combustion. *Journal of American Ceramic Society* 132: 15–19.

Karamberi, A. and Moutsatsou, A. 2012. Vitrification of lignite fly ash and metal slags for the production of glass and glass ceramics. *China Particuology* 4(5): 250–253.

Karimi, E., Gomez, A., Kycia, S.W., and Schlaf, M. 2010. Thermal decomposition of acetic and formic acid catalyzed by red mud—implications for potential use of red mud as a pyrolysis bio-oil upgrading catalyst. *Energy & Fuels* 24: 2747–2757.

Karimian, N., Kalbasi, M., and Hajrasuliha, S. 2012. Effect of converter sludge, and its mixtures with organic matter, elemental sulfur and sulfuric acid on availability of iron, phosphorus and manganese of 3 calcareous soils from central Iran. *African Journal of Agricultural Research* 7(4): 568–576.

Kashiwaya,Y., Akiyama, T., and In-Nami, Y. 2010. Latent heat of amorphous slags and their utilization as a high temperature PCM. *ISIJ International* 50(9): 1259–1264.

Kevorkijan, V.M. 1999. The quality of aluminum dross particles and cost-effective reinforcement for structural aluminum-based composites. *Composites Science and Technology* 59: 1745–1751.

Khan, Z.A., Malkawi, R.H., Al-Ofi, K.A., and Khan, N. 2002. Review of steel slag utilization in Saudi Arabia. The 6th Saudi Engineering Conference, December 14–17, 2002, KFUPM, Dhahran.

Khater, G.A. 2002. The use of Saudi slag for the production of glass-ceramic materials. *Ceramics International* 28: 59–67.

Kim, E.H., Yim, S.B., Jung, H.C., and Lee, E.J. 2006. Hydroxyapatite crystallization from a highly concentrated phosphate solution using powdered converter slag as a seed material. *Journal of Hazardous Materials B* 136: 690–697.

Kim, J., Biswas, K., Jhon, K.W., Jeong, S.Y., and Ahn, W.S. 2009. Synthesis of AlPO4-5 and CrAPO-5 using aluminum dross. *Journal of Hazardous Materials* 169: 919–925.

Kitamura, S., Shibata, H., Kim, S., Teradoko, T., Maruoka, N., and Yamaguchi, K. 2011. European environmental policy and its influence on the use of slag products. Second International Slag Valorisation Symposium, The Transition to Sustainable Materials Management, April 18–20, 2011, Leuven, Belgium, pp. 201–213.

Klauber, C., Gräfe, M., and Power, G. 2011. Bauxite residue issues: II. Options for residue utilization. *Hydrometallurgy* 108: 11–32.

Kolesnikova, M.P., Saigofarov, S.S., Nikonenko, E.A., Kalinichenko, I.I., Kochneva, T.P., and Surkova, N.A. 1998. The use of red mud for brick coloring. *Glass and Ceramics* 55: 70–71.

Kopkova, E.K., Gromov, P.B., and Shchelokova, E.A. 2011. Decomposition of converter copper-nickel slag in solutions of sulfuric acid. *Theoretical Foundations of Chemical Engineering* 45(4): 505–510.

Kriskova, L., Pontikes, Y., Cizer, O., Mertens, G., Veulemans, W., Geysen, D., Jones, P.T., Vandewalle, L., Balen, K.V., and Blanpain, B. 2012. Effect of mechanical activation on the hydraulic properties of stainless steel slags. *Cement and Concrete Research* 42: 778–788.

Kumar, R.V. and Liu, J. 2011. Opportunity for using steelmaking/EAF or BOF laden dust to desulphurise hot metal. Second International Slag Valorisation Symposium, The Transition to Sustainable Materials Management, April 18–20, 2011, Leuven, Belgium, pp. 287–298.

Kumar, S., Kumar, R., and Bandopadhyay, A. 2006. Innovative methodologies for the utilisation of wastes from metallurgical and allied industries. *Resources, Conservation and Recycling* 48: 301–314.

Kuwahara, Y., Tsuji, K., Ohmichi, T., Kamegawa, T., Mori, K., and Yamashita, H. 2012a. Transesterifications using a hydrocalumite synthesized from waste slag: an economical and ecological route for biofuel production. *Catalysis Science & Technology* 2: 1842–1851.

Kuwahara, Y., Tsuji, K., Ohmichi, T., Kamegawa, T., Mori, K., and Yamashita, H. 2012b. Waste-slag hydrocalumite and derivatives as heterogeneous base catalysts. *ChemSusChem* 5: 1523–1532.

Lazzaro, G., Eltrudis, M., and Pranovi, F. 1994. Recycling of aluminium dross in electrolytic pots. *Resources, Conservation and Recycling* 10: 153–159.

Legemza, J. 2004. The possibilities of utilizing dust and sludge from steel industry. *Acta Metallurgica Slovaca* 10(2): 80–87.

Lei, L., Hua W., and Hu, J. 2010. The study of recovering iron by melting reduction from copper slag. International Conference on Digital Manufacturing & Automation, December 18–20, 2010, Chang Sha, China, pp. 823–827.

Li, G. and Ni, H. 2011. Recent progress of hot stage processing for steelmaking slags in China considering stability and heat recovery. Second International Slag Valorisation Symposium, The Transition to Sustainable Materials Management, April 18–20, 2011, Leuven, Belgium, pp. 253–261.

Li, J., Wang, J., Chen, H., and Sun, B. 2012. Microstructure observation of β-sialon-15R ceramics synthesized from aluminum dross. *Ceramics International* 38: 3075–3080.

Li, Y., Liu, C., Luan, Z., Peng, X., Zhu, C., Chen, Z., Zhang, Z., Fan, J., and Jia, Z. 2006. Phosphate removal from aqueous solutions using raw and activated red mud and fly ash. *Journal of Hazardous Materials B* 137: 374–383.

Li, Y., Papangelakis, V.G., and Perederiy, I., 2009. High pressure oxidative acid leaching of nickel smelter slag: characterization of feed and residue. *Hydrometallurgy* 97: 185–193.

Li, Y., Perederiy, I., and Papangelakis, V.G. 2008. Cleaning of waste smelter slags and recovery of valuable metals by pressure oxidative leaching. *Journal of Hazardous Materials* 152: 607–615.

Lim, T.T. and Chu, J. 2006. Assessment of the use of spent copper slag for land reclamation. *Waste Management and Research* 24: 67–73.

Lindsay, R.D. 1995. Process for recovery of free aluminum from aluminum dross or aluminum scrap using plasma energy with oxygen second stage treatment. US patent 5,447,548.

Lisbona, D.F., Somerfield, C., and Steel, K.M. 2012. Leaching of spent pot-lining with aluminum anodizing wastewaters: fluoride extraction and thermodynamic modeling of aqueous speciation. *Industrial & Engineering Chemistry Research* 51: 8366–8377.

Liu, C., Shi, P., Zhang, D., and Jiang, M. 2007. Development of glass ceramics made from ferrous tailings and slag in China. *Journal of Iron and Steel Research International* 14(2): 73–78.

Liu, Q., Xin, R., Li, C., Xu, C., and Yang, J. 2013. Application of red mud as a basic catalyst for biodiesel production. *Journal of Environmental Sciences* 25:823–829.

Liu, S.Y., Gao, J., Yang, Y.J., Yang, Y.C., and Ye, Z.X. 2010. Adsorption intrinsic kinetics and isotherms of lead ions on steel slag. *Journal of Hazardous Materials* 173: 558–562.

Liu, Y., Naidu, R., and Ming, H. 2011. Red mud as an amendment for pollutants in solid and liquid phases. *Geoderma* 163: 1–12.

López, A., de Marco, I., Caballero, B.M., Laresgoiti, M.F., Adrados, A., and Aranzabal, A. 2011. Catalytic pyrolysis of plastic wastes with two different types of catalysts: ZSM-5 zeolite and red mud. *Applied Catalysis B: Environmental* 104: 211–219.

Lopez, F.A., Perez, C., Sainz, E., and Alonso, M. 1995. Adsorption of Pb^{2+} on blast furnace sludge. *Journal of Chemical Technology and Biotechnology* 62: 200–206.

Lucheva, B., Tsonev, T., and Petkov, R. 2005. Non-waste aluminium dross recycling. *Journal of the University of Chemical Technology and Metallurgy* 40: 335–338.

Ma, Y., Si, C., and Lin, C. 2012. Comparison of copper scavenging capacity between two different red mud types. *Materials* 5: 1708–1721.

Manfredi, O., Wuth, W., and Bohlinger, I. 1997. Characterizing the physical and chemical properties of aluminum dross. *JOM* 49: 48–51.

Matusewicz, R. and Roberts, J.S. 2002. Ausmelt technology smelting unit for the processing of spent pot lining (SPL) at Portland. *Aluminium* 78: 1–2.

Mazumder, B. and Devi, S.R. 2008. Adsorption of oils, heavy metals and dyes by recovered carbon powder from spent pot liner of aluminum smelter plant. *Journal of Environmental Science and Engineering* 50: 203–206.

Meor, Y.M.S., Masliana, M., and Wilfred, P. 2010. Effect of fractional precipitation on quality of nanostructured alumina produced from black aluminium dross waste. *Advanced Materials Research* 173: 24–28.

Mihok, I., Seilerová, K., and Baricová, D. 2004. Recycling of steelmaking slag from electric arc furnace. *Archives of Foundry* 4 (13): 165–169.

Mihok, L., Demeter, P., Baricova, D., and Seilerova, K. 2006. Utilization of ironmaking and steelmaking slags. *Metalurgija* 45(3): 163–168.

Miksa, D., Homsak, M., and Samec, N. 2003. Spent potlining utilization possibilities. *Waste Management Research* 21: 467–473.

Monshi, A. and Asgarani, M.A. 1999. Producing Portland cement from iron and steel slags and limestone. *Cement and Concrete Research* 29: 1373–1377.

Morita, K., Guo, M., and Oka, N. 2002. Nobuo sanoresurrection of the iron and phosphorus resource in steel-making slag. *Journal of Material Cycles and Waste Management* 4: 93–101.

Morrison, C. and Richardson, D. 2004. Re-use of zinc smelting furnace slag in concrete. *Engineering Sustainability* 157(ES4): 213–218.

Motz, H. and Geiseler, J. 2001. Products of steel slags: an opportunity to save natural resources. *Waste Management* 21: 285–293.

Moxnes, B., Gikling, H., Kvande, H., Rolseth, S., and Straumsheim, K. 2003. Addition of refractories from spent potlining to alumina reduction cells to produce Al-Si alloys. In: *Light Metals* Crepeau, P. N., Ed., TMS (The Minerals, Metals & Materials Society), Warrendale, PA, pp. 329–334.

Mudersbach, D., Drissen, P., and Motz, H. 2011. Improved slag qualities by liquid slag treatment. Second International Slag Valorisation Symposium, The Transition to Sustainable Materials Management, April 18–20, 2011, Leuven, Belgium, pp. 299–311.

Mukhopadhyay, J., Ramana, Y.V., and Singh, U. 2004. Extraction of value added products from aluminium dross material to achieve zero waste. In: *Light Metals,* Tabereaux, A.T., Ed., TMS (The Minerals, Metals & Materials Society), Warrendale, PA, p. 4.

Mulopo, J., Mashego, M., and Zvimba, J.N. 2012. Recovery of calcium carbonate from steel-making slag and utilization for acid mine drainage pre-treatment. *Water Science and Technology,* in press.

Muravyov, M.I., Fomchenko, N.V., Usoltsev, A.V., Vasilyev, E.A., and Kondrat'eva, T.F. 2012. Leaching of copper and zinc from copper converter slag flotation tailings using H_2SO_4 and biologically generated $Fe_2(SO_4)_3$. *Hydrometallurgy* 119–120: 40–46.

Murayama, N., Okajima, N., Yamaoka, S., Yamamoto, H., and Shibata, J. 2006. Hydrothermal synthesis of AlPO4-5 type zeolitic materials by using aluminum dross as a raw material. *Journal of the European Ceramic Society* 26: 459–462.

Murayama, N., Arimura, K., Okajima, N., and Shibata, J. 2009. Effect of structure-directing agent on AlPO4-n synthesis from aluminum dross. *International Journal of Mineral Processing* 93: 110–114.

Murayama, N., Maekawa, I., Ushiro, H., Miyoshi, T., Shibata, J., and Valix, M. 2012. Synthesis of various layered double hydroxides using aluminum dross generated in aluminum recycling process. *International Journal of Mineral Processing* 110–111: 46–52.

Nikitin, L.D. et al. 2001. Use of aluminium production wastes in the charge of blast furnaces at the West Siberian Metallurgical Plant. *Chernaya Metalllurgiya, Byulleten Nauchno-Tekhnicheskoi I Ekonomicheskoi Informatsii* 11, 3336 1, 82–86.

Nippon Slag Association. 2012a. http://www.slg.jp/e/images/Amounts%20of%20Blast%20 Furnace%20Slag.pdf.

Nippon Slag Association. 2012b. http://www.slg.jp/e/images/Amounts%20of%20Steel%20 Slag.pdf.

O'Connor, W.K., Turner, P.C., and Addison, G.W. 2002. Method for processing aluminum spent potliner in a graphite electrode ARC furnace. US Patent 6498282.

OECD Materials Case Study 2. 2010. Aluminium Working Document, OECD Environment Directorate, Global Forum on Environment, Focusing on Sustainable Materials Management, October 25–27, 2010, Mechelen, Belgium.

Oguz, E. 2004. Removal of phosphate from aqueous solution with blast furnace slag. *Journal of Hazardous Materials B* 114: 131–137.

Oliveira, D.R.C. and Rossi, C.R.C. 2012. Concretes with red mud coarse aggregates. *Materials Research* 15: 333–340.

Oliveira, G.E. and Holanda, J.N.F. 2004. Use of mix of clay/solid waste from steel works for civil construction materials. *Waste Management Research* 22: 358–363.

Oliviera, A.A.S., Texeira, I.F., Ribeiro, L.P., Tristao, J.C., Dias, A., and Lago, R.M. 2010. Magnetic amphiphilic composites based on carbon nanotubes and nanofibers grown on an inorganic matrix: effect on water-oil interfaces. *Journal of the Brazilian Chemical Society* 21: 2184–2188.

Onisei, S., Pontikes, Y., Gerven, T.V., Angelopoulos, G.N., Velea, T., Predica, V., and Moldovan, P. 2012. Synthesis of inorganic polymers using fly ash and primary lead slag. *Journal of Hazardous Materials* 205–206: 101–110.

Ordóñez, S., Sastre, H., and Díez, F.V. 2001. Catalytic hydrodechlorination of tetrachloroethylene over red mud. *Journal of Hazardous Materials* 81: 103–114.

Ortiz, N., Pires, M.A.F., and Ressiani, J.C. 2001. Use of steel converter slag as nickel adsorber to wastewater treatment. *Waste Management* 21: 631–635.

Øye, H. 1994. Treatment of spent potlining in aluminium electrolysis, a major engineering and environment challenge. *Energeia* 5(1): 1–6. http://www.caer.uky.edu/energeia/PDF/vol5_1.pdf.

Palmer, S.J., Nothling, M., Bakon, K.H., and Frost. R.L. 2010. Thermally activated seawater neutralised red mud used for removal of arsenate, vanadate and molybdate from aqueous solutions. *Journal of Colloid and Interface Science* 342: 147–154.

Park, D., Lim, S.R., Lee, H.W., and Park, J.M. 2008. Mechanism and kinetics of Cr(VI) reduction by waste slag generated from iron making industry. *Hydrometallurgy* 93: 72–75.

Pawlek, R.P. 2012. Spent potlining: an update. In: *Light Metals* Suarez, C.E., Ed., TMS (The Minerals, Metals & Materials Society), Warrendale, PA, pp. 1313–1317.

Peng, C., Zhi, X., Gu, M., He, R., and Guo, Z. 2011. Recycling of some steel industrial solid wastes for high-value material applications. Second International Slag Valorisation Symposium, The Transition to Sustainable Materials Management, April 18–20, 2011, Leuven, Belgium, pp. 341–349.

Piga, L., Pochetti, F., and Stoppa, L. 1993. Recovering metals from red mud generated during alumina production. *JOM* 45: 54–59.

Plunkert, P.A. 2006. Aluminum recycling in the United States in 2000. *U.S. Geological Survey Circular* 1196–W.

Pontikes, Y., Rathossi, C., Nikolopoulos, P., Angelopoulos, G.N., Jayaseelan, D.D., and Lee, W.E. 2009. Effect of firing temperature and atmosphere on sintering of ceramics made from Bayer process bauxite residue. *Ceramics International* 35: 401–407.

Pontikes, Y., Kriskova, L., Wang, X., Geysen, D., Arnout, S., Nagels, E., Cizer, O., Gerven, T.V., Elsen, J., Guo, M., Jones, P.T., and Blanpain, B. 2011. Additions of industrial residues for hot stage engineering of stainless steel slags. Second International Slag Valorisation Symposium, The Transition to Sustainable Materials Management, April 18–20, 2011, Leuven, Belgium, pp. 313–326.

Power, G., Gräfe, M., and Klauber, C. 2011. Bauxite residue issues: I. Current management, disposal and storage practices. *Hydrometallurgy* 108: 33–45.

Pradhan, J., Das, S.N., and Thakur, R.S. 1999. Adsorption of hexavalent chromium from aqueous solution by using activated red mud. *Journal of Colloid and Interface Science* 217: 137–141.

Pratt, K.C. and Christoverson, V. 1982. Hydrogenation of a model hydrogen-donor system using activated red mud catalyst. *Fuel* 61: 460–462.

Proshkin, A.V., Pavlov, V.F., and Khokhlov, A.I. 2002. Recycling of the reduction cells used lining into thermal insulating materials. Proceedings VIII Aluminium of Siberia Symposium, Sept. 10–12, 2002, Krasnoyarsk, pp. 105–111.

Puertas, F. and Jimenez, A.F. 2003. Mineralogical and microstructural characterization of alkali-activated fly ash/slag pastes. *Cement & Concrete Composites* 25: 287–292.

Pulford, I.D., Hargreaves, J.S.J., Ďurišovà, J., Kramulovam, B., Girard, C., Balakrishnan, M., Batra, V.S., and Rico, J.L. 2012. Carbonised red mud—a new water product made from a waste material. *Journal of Environmental Management* 100: 59–64.

Purwanto, H. and Akiyama, T. 2006. Hydrogen production from biogas using hot slag. *International Journal of Hydrogen Energy* 31: 491–495.

Qin, S. and Wu, B. 2011. Reducing the radiation dose of red mud to environmentally acceptable levels as an example of novel ceramic materials. *Green Chemistry* 13: 2423–2427.

Reuter, M., Xiao, Y., and Boin, U. 2004. Recycling and environmental issues of metallurgical slags and salt fluxes. VII International Conference on Molten Slags Fluxes and Salts, The South African Institute of Mining and Metallurgy, January 25–28, 2004, Johannesburg, South Africa, pp. 349–356.

Rio Tinto. 2011. http://www.riotintoalcan.com/documents/Reports_July2011_RioTintoCanada Brochure_EN.pdf.

Rusen, A., Geveci, A., and Topkaya, Y.A. 2012. Minimization of copper losses to slag in matte smelting by colemanite addition. *Solid State Sciences* 14: 1702–1704.

Ruyters, S., Mertens, J., Vassilieva, E., Dehandschutter, B., Polfijn, A., and Smolders, E. 2011. The red mud accident in Ajka (Hungary): plant toxicity and trace metal bioavailability in red mud contaminated soil. *Environmental Science & Technology* 45: 1616–1622.

Saikia, N., Cornelis, G., Cizer, O., Vandecasteele, C., Gemert D.V., Balen, K.V., and Gerven, T.V. 2012. Use of Pb blast furnace slag as a partial substitute for fine aggregate in cement mortar. *Journal of Material Cycles Waste Management* 14: 102–112.

Santona, L., Castaldi, P., and Melis, P. 2006. Evaluation of the interaction mechanisms between red muds and heavy metals. *Journal of Hazardous Materials B* 136: 324–329.

Santos, R.M., Ling, D., Sarvaramini, A., Guo, G., Elsen, J., Larachi, F., Beaudoin, G., Blanpain, B., and Gerven, T.V. 2012. Stabilization of basic oxygen furnace slag by hot-stage carbonation treatment. *Chemical Engineering Journal* 203: 239–250.

Sarkar, R., Singh, N., and Das, S.K. 2010. Utilization of steel melting electric arc furnace slag for development of vitreous ceramic tiles. *Bulletin of Material Science* 33(3): 293–298.

Sato, S., Yoshikawa, T., Nakamoto, M., Tanaka, T., and Ikeda, J. 2008. Application of hydrothermal treatment on BF slag and waste glass for preparing lubricant materials in high strain rolling for ultrafine-grained steel production. *ISIJ International* 48(2): 245–250.

Schwarz, H.G. 2004. Aluminum production and energy. *Encyclopedia of Energy*, pp. 81–95. Elsevier.

Shen, H. and Forssberg, E. 2003. An overview of recovery of metals from slags. *Waste Management* 23: 933–949.

Shi, C. 2004. Steel slag-its production, processing, characteristics, and cementitious properties. *Journal of Materials in Civil Engineering* 16: 230–236.

Shi, C., Meyer, C., and Behnood, A. 2008. Utilization of copper slag in cement and concrete. *Resources, Conservation and Recycling* 52: 1115–1120.

Shi, C. and Qian, J. 2000. High performance cementing materials from industrial slags-a review. *Resources, Conservation and Recycling* 29: 195–207.

Shi, Z., Li, W., Hu, X., Ren, B., Gao, B., and Wang, Z. 2012. Recovery of carbon and cryolite from spent pot lining of aluminium reduction cells by chemical leaching. *Transactions of Nonferrous Metals Society of China* 22: 222–227.

Shih, P.H., Wu, Z.Z., and Chiang, H.L. 2004. Characteristics of bricks made from waste steel slag. *Waste Management* 24: 1043–1047.

Shilton, A.N., Elmetri, I., Drizo, A., Pratt, S., Haverkamp, R G., and Bilby, S.C. 2006. Phosphorus removal by an "active" slag filter—a decade of full scale experience. *Water Research* 40: 113–118.

Shinzato, M.C. and Hypolito, R. 2005. Solid waste from aluminum recycling process: characterisation and reuse of its economically valuable constituents. *Waste Management* 25: 37–46.

Silveira, B.I., Dantas, A.E.M., Blasques, J.E.M., and Santos R.K.P. 2003. Effectiveness of cement-based systems for stabilization and solidification of spent pot liner inorganic fraction. *Journal of Hazardous Materials B* 98: 183–190.

Singh, A.K.P. and Raju, M.T. 2011. Recycling of basic oxygen furnace (BOF) sludge in iron and steel works. *International Journal of Environmental Technology and Management* 14: 19–32.

Snars, K. and Gilkes, R.J. 2009. Evaluation of bauxite residues (red muds) of different origins for environmental applications. *Applied Clay Science* 46: 13–20.

Sofilić, T., Sofilić, U., and Brnardić, I. 2012. The significance of iron and steel slag as by-product for utilization in road construction. 12th International Foundrymen Conference, Sustainable Development in Foundry Materials and Technologies, May 24–25, 2012, Opatija, Croatia.

Somlai, J., Jobbagy, V., Kovaks, J., Tarjan, S., and Kovacs, T. 2008. Radiological aspects of the usability of red mud as building additive. *Journal of Hazardous Materials* 150: 541–545.

Sorlini, S., Sanzeni, A., and Rondi, L. 2012. Reuse of steel slag in bituminous paving mixtures. *Journal of Hazardous Materials* 209–210: 84–91.

Su, F., Lampinen, H.O., and Robinson, R. 2004. Recycling of sludge and dust to the BOF converter by cold bonded pelletizing. *ISIJ International* 44(4): 770–776.

Summers, R.N. and Pech, J.D. 1997. Nutrient and metal content of water, sediment and soils amended with Bauxite residue in the catchment of the Peel Inlet and Harvey Estuary, Western Australia. *Agriculture, Ecosystems and Environment* 64: 219–232.

Sushil, S., Alabdulrahman, A.M., Balakrishnan, M., Batra, V.S., Blackley, R.A., Clapp, J., Hargreaves, J.S.J., Monaghan, A., Pulford, I.D., Rico, J.L., and Zhou, W. 2010. Carbon deposition and phase transformations in red mud on exposure to methane. *Journal of Hazardous Materials* 180: 409–418.

Sushil, S. and Batra, V.S. 2008. Catalytic applications of red mud, an aluminium industry waste: a review. *Applied Catalysis B: Environmental* 81: 64–77.

Tenorio, J.A.S. and Espinosa, D.C.R. 2002. Effect of salt/oxide interaction on the process of aluminum recycling. *Journal of Light Metals*. 2: 89–93.

Tenório, J.A.S. and Espinosa, D.C.R. 2003. Aluminum recycling. In: *Handbook of Aluminum: Production and Materials Manufacturing*, Vol. 2, Totten, G.E. and Mackenzie, D.S., Eds., Marcel Dekker, New York, pp. 115–153.

Tetronics. 2011. Treatment of aluminium spent potliner waste. http://www.tetronics.com/File/SpentPotlinerDatasheet.pdf.

The European Slag Association. 2012. Position paper on the status of ferrous slag. http://www.euroslag.org/fileadmin/_media/images/Status_of_slag/Position_Paper_April_2012.pdf.

Tossavainen, M., Engstrom, F., Yang, Q., Menad, N., Larsson, M.L., and Bjorkman, B. 2007. Characteristics of steel slag under different cooling conditions. *Waste Management* 27: 1335–1344.

Tsakiridis, P.E. 2012. Aluminium salt slag characterization and utilization—a review. *Journal of Hazardous Materials* 217–218: 1–10.

Unlu, N. and Drouet, M.G. 2002. Comparison of salt-free aluminum dross treatment processes. *Resources, Conservation and Recycling* 36: 61–72.

van Oss, H.G. 2012. Iron and steel slag. U.S. Geological Survey, Mineral Commodity Summaries, http://minerals.usgs.gov/minerals/pubs/commodity/iron_&_steel_slag/mcs-2012-fesla.pdf.

Venancio, L.C.A., Souza, J.A.S., Macedo, E.N., Quaresma, J.N.N., and Paiva, A.E.M. 2010. Residues recycling: reduction costs and helping the environment. *JOM* 62: 41–45.

Vereš, J., Jakabský, S., and Lovás, M. 2011. Zinc recovery from iron and steel making wastes by conventional and microwave assisted leaching. *Acta Montanistica Slovaca Ročník* 16(3): 185–191.

Vieira, C.M.F., Andrade, P.M., Maciel, G.S., Vernilli Jr., F., and Monteiro, S.N. 2006. Incorporation of fine steel sludge waste into red ceramic. *Materials Science and Engineering A* 427: 142–147.

Vlček, J., Tomková, V., Ovčačíková, H., Martinec, P., Volková, A., Topinková, M., Matějka, V., Ovčačík, F., and Michnová, M. 2012. Slag from production of pig iron and steel making and possibilities of their utilization. 21st International Conference on Metallurgy and Materials, May 23–25, 2012, Brno, Czech Republic.

von Krüger, P. 2011. Use of spent potlining (SPL) in ferro silico manganese smelting. In: *Light Metals*, Lindsay, S.J., Ed., TMS (The Minerals, Metals & Materials Society), Warrendale, PA, pp. 275–280.

Waligora, J., Bulteel, D., Degrugilliers, P., Damidot, D., Potdevin, J.L., and Measson, M. 2010. Chemical and mineralogical characterizations of LD converter steel slags: a multi-analytical techniques approach. *Material Characterization* 61: 39–48.

Wang, W., Xue, Z., Ma, G., Xiao, H., Guo, X., and Xing, L. 2010. Producing iron nuggets with steel making wastes. 4th International Conference on Bioinformatics and Biomedical Engineering (ICBBE), June 18–20, 2010, Chengdu, China, pp. 1–4.

Wang, G. and Emery, J. 2004. Technology of slag utilization in highway construction. Environmental Benefits of In-situ Material Recycling and Strengthening Session, Annual Conference of the Transportation Association of Canada, September 19–22, 2004, Québec City, Québec.

Wang, M., Wang X., He, Y., Lou, T., and Sui, Z. 2008. Isothermal precipitation and growth process of perovskite phase in oxidized titanium bearing slag. *Transactions of Nonferrous Metals Society of China* 18: 459–462.

Wang, S., Ang, H.M., and Tadé, M.O. 2008. Novel applications of red mud as coagulant, adsorbent and catalyst for environmentally benign processes. *Chemosphere* 72: 1621–1635.

Wang, X., Geysen, D., Padilla, S.V.T., D'Hoker, N., Huang, S., Jones, P.T., Gerven, T.V., and Blanpain, B. 2011. Fayalite based slags: metal recovery and utilization. Second International Slag Valorisation Symposium, The Transition to Sustainable Materials Management, April 18–20, 2011 Leuven, Belgium.

Wang, Z., Ni, W., Li, K., Huang, X., and Zhu, L. 2011. Crystallization characteristics of iron-rich glass ceramics prepared from nickel slag and blast furnace slag. *International Journal of Minerals, Metallurgy and Materials* 18(4): 455–459.

Waste Management World. 2011. Aluminium dross recycled into fertilizer in New Zealand. http://www.waste-management-world.com/index/display/article-display.articles.waste-management-world.recycling.2011.10b.Aluminium_Dross_Recycled_into_Fertilizer_in_New_Zealand.QP129867.dcmp = rss.page = 1.html.

World Steel Association. 2011. Steel production 2001, http://www.worldsteel.org/statistics/statistics-archive/2011-steel-production.html; iron production 2011, http://www.worldsteel.org/statistics/statistics-archive/2011-iron-production.html.

Wu Z. and Zongshu, Z. 2005. Utilization of ferrous metallurgical slag as resource materials in agriculture. 7th World Congress on Recovery, Recycling and Re-integration, September 25–29, 2005, Beijing, China.

Wu, S., Xue, Y., Ye, Q., and Chen, Y. 2007. Utilization of steel slag as aggregates for stone mastic asphalt (SMA) mixtures. *Building and Environment* 42: 2580–2585.

Wu, Z.J., Jiang, B.F., Liu, W.M., Cao, F.B., Wu, X.R., and Li, L.S. 2011. Selective recovery of valuable components from converter steel slag for preparing multidoped $FePO_4$. *Industrial & Engineering Chemistry Research* 50: 13778–13788.

Wu, Z.J., Zhou, Y., Su, S.H, Gao, Z.F., Wu, X.R, and Li, L.S. 2012. A novel conversion of converter sludge into amorphous multi-doped $FePO_4$ cathode material for lithium ion batteries. *Scripta Materialia* 67: 221–224.

Xiong, J., He, Z., Mahmood, Q., Liu, D., Yang, X., and Islam, E. 2008. Phosphate removal from solution using steel slag through magnetic separation. *Journal of Hazardous Materials* 152: 211–215.

Xue, Y., Hou, H., and Zhu, S. 2009. Competitive adsorption of copper(II), cadmium(II), lead(II) and zinc(II) onto basic oxygen furnace slag. *Journal of Hazardous Materials* 162: 391–401.

Yamada, K., Harato, T., and Shiozaki, Y. 1980. Process for the removal of sulfur oxides from exhaust gases using slurry of red mud containing calcium ion. US Patent 4,222,992.

Yang, Z., Rui-lin, M., Wang-dong, N., and Hui, W. 2010. Selective leaching of base metals from copper smelter slag. *Hydrometallurgy* 103: 25–29.

Yerushalmi, D. and Sarko, L. 1995. Method of recycling aluminium dross. US patent 5,424,260.

Yildirim, I.Z. and Prezzi, M. 2011. Chemical, mineralogical, and morphological properties of steel slag. *Advances in Civil Engineering*, 2011, 1–13.

Yoshimura, H.N., Abreu, A.P., Molisani, A.L., de Camargo, A.C., Portela, J.C.S., and Narita, N.E. 2008. Evaluation of aluminum dross waste as raw material for refractories. *Ceramics International* 34: 581–591.

Zhai, X., Li, N., Zhang, X., Fu, Y., and Jiang, L. 2011. Recovery of cobalt from converter slag of Chambishi Copper Smelter using reduction smelting process. *Transactions of Nonferrous Metals Society of China* 21: 2117–2121.

Zhang, X., Matsuura, H., and Tsukihashi, F. 2011. Utilisation of steelmaking slag for improvement of coastal environment. Second International Slag Valorisation Symposium, The Transition to Sustainable Materials Management, April 18–20, 2011, Leuven, Belgium, pp. 271–278.

3 Coal Combustion Waste Materials

J. Groppo

CONTENTS

3.1   COAL COMBUSTION

3.1.1   Power Production from Coal Combustion

Coal is the most widely used fuel for electrical generation in the world, responsible for more than 40% of the world's electricity generation (IEA, 2011). Although the technologies used to accomplish this task are varied, the principles are basically similar. Coal is burned in a boiler or furnace to generate heat in order to produce steam. The steam is generated by circulating water through a network of tubes located throughout the boiler. Superheated steam is then directed into a turbine where it expands to turn the turbine shaft, which is coupled to the shaft of a generator, which turns to produce the electrical current. Condensed water and low-pressure steam are recirculated back to the boiler where they are reheated to produce additional steam in a continuous closed heating loop.

3.1.2   Coal Mineralogy and Chemistry

Since coal is the primary component of most of the by-products produced during combustion, some understanding of coal mineralogy and chemistry is a prerequisite to understanding the characteristics of coal ash. Coal is the general term applied to numerous organic minerals of varying composition and properties; however, all of these minerals are enriched in carbon and are generally black in color. All coals originate from the slow decomposition and chemical conversion of immense masses of organic material (Hower and Parekh, 1991). Coal formation is a continuous process, and as the chemical conversion proceeds with time, physical and chemical properties of the organic matter change with a general increase in carbon content, referred to as rank.

Specific coal rank classifications vary between countries, but in general, formation begins with peat, and sequentially progresses to lignite, subbituminous coal, bituminous coal, and anthracite. As this progression proceeds, several important compositional transformations occur. As previously mentioned, carbon content increases with coal rank. This increase occurs as volatile matter is converted into carbon; thus, volatile matter decreases with rank. Another important change is that moisture content decreases with rank. This moisture is not free or surface water, but rather it is moisture within the chemical structure of the coal itself. Peat can contain as much as 75% chemically bound water, which decreases to 35% in lignite, further decreasing to 25% in subbituminous coal, and to less than 10% moisture in bituminous coal (Teichmüller and Teichmüller, 1982). Moisture content is particularly important when considering thermal properties of coal because higher moisture content results in lower heating value.

In addition to carbon, volatile matter, and moisture, inorganic material is also found in coal. The source of the inorganic material is weathering and erosion of associated minerals that are deposited along with organic matter throughout the coal formation process. A variety of minerals can be found in coal in varying concentrations, depending on the depositional environment, as shown in Table 3.1. Similarly, trace elements are also found in coal in varying concentrations (Table 3.2). Coal preparation or cleaning with physical beneficiation methods can

TABLE 3.1

Classification of Common Minerals Found in Coal

Shale Group (Group M)	Accessory Minerals
Muscovite $(KAl_2(AlSiO_3O_{10})(OH)_2)$	Sphalerite (ZnS)
Hydromuscovite	Feldspar $(K,Na)_2O \cdot Al_2O_3 \cdot 6SiO_2$
Illite $(K(MgAl,Si)(Al,Si_3)O_{10}(OH)_8)$	Garnet $(3CaO \cdot Al_2O_3 \cdot 3SiO_2)$
Bravaisite	Hornblende $(CaO \cdot 3FeO \cdot 4SiO_2)$
Montmorillonite $(MgAl)_3(Si_4O_{10})_3(OH)_{10} \cdot 12H_2O$	Gypsum $(CaSO_4 \cdot 2H_2O)$
Kaolin group (Group K)	Apatite $(9CaO \cdot 3P_2O_5 \cdot CaF_2)$
Kaolinite $(Al_2SI_2O_5)OH)_4)$	Zircon $(ZrSiO_4)$
Levisite	Epidote $(4CaO \cdot 3Al_2O_3 \cdot 6SiO_2 \cdot H_2O)$
Metahalloysite	Biotite $(K_2O \cdot MgO \cdot Al_2O_3 \cdot 3SiO_2 \cdot H_2O)$
Sulfide group (Group S)	Augite $(CaO \cdot MgO \cdot 2SiO_2)$
Pyrite (FeS_2)	Prochlorite $(2FeO \cdot 2MgO \cdot Al_2O_3 \cdot 2SiO_2 \cdot 2H_2O$
Marcasite (FeS_2)	Chlorite $(Mg,Fe,Al)_8(Si,Al)_4O_{10}(OH)_8$
Carbonate group (Group C)	Diaspore $(Al_2O_3 \cdot H_2O)$
Ankerite $CaCO_3 \cdot (Mg, Fe, Mn)CO_3$	Lepidocrocite $(Fe_2O_3 \cdot H_2O)$
Calcite $(CaCO_3)$	Barite $(BaSO_4)$
Siderite $(FeCO_3)$	Kyanite $(Al_2O_3 \cdot SiO_2)$
Chloride Group (Group O)	Staurolite $(2FeO \cdot 5Al_2O_3 \cdot 4SiO_2 \cdot H_2O)$
Sylvite (KCl)	Topaz $(AlF)_2SiO_4$
Halite (NaCl)	Tourmaline $(H_9Al_3(BOH)_2Si_4O_{19}$
Oxide group (Group O)	Pyrophyllite $(Al_2Si_4O_{10}(OH)_2)$
Quartz (SiO_2)	Penninite $(5MgO \ Al_2O_3 \ 3SiO_2 \ 2H_2O)$
Hematite (Fe_2O_3)	
Magnetite (Fe_3O_4)	

Source: Nelson, 1953; Spackman and Moses, 1961.

remove some of the inorganic mineral and trace elements associated with coal; however, the extent of removal is dependent on liberation of the inorganic species. If minerals and trace elements are associated with the organic matrix, physical cleaning will do little to remove them prior to combustion. As a result, the inorganic elements contained in the coal will be effectively concentrated during combustion as the carbon and volatile matter are removed by combustion. The implication is that these elements will be concentrated in the combustion by-products in either gaseous or solid form.

3.1.3 COAL UTILIZATION

As previously stated, coal is the predominant fuel used to generate most of the world's electricity, with an estimated production of hard coal and lignite of 7229 Mt in 2010 (IEA, 2011). The majority of global coal production (hard coal and brown coal/lignite) was used for electrical generation while the remainder was used for

TABLE 3.2

Mean Concentration (ppm) of Trace Elements in U.S. Coals

Element	Eastern U.S.	Illinois Basin	Western U.S.
Ag	0.02	0.03	0.03
As	25	14	2.3
B	42	110	56
Ba	200	100	500
Be	1.3	1.7	0.46
Br	12	13	4.7
Cd	0.24	2.2	0.18
Ce	25	14	11
Co	9.8	7.3	1.8
Cr	20	18	9
Cs	2	1.4	0.42
Cu	18	14	10
Dy	2.3	1.1	0.63
Eu	0.52	0.26	0.2
F	89	67	62
Ga	5.7	3.2	2.5
Ge	1.6	6.9	0.91
Hf	1.2	0.54	0.78
Hg	0.2	0.2	0.09
I	1.7	1.7	0.52
In	0.23	0.16	0.1
La	15	6.8	5.2
Lu	0.22	0.09	0.07
Mn	18	53	49
Mo	4.6	8.1	2.1
Ni	15	21	5
P	150	64	130
Pb	5.9	32	3.4
Rb	22	19	4.6
Sb	1.6	1.3	0.58
Sc	5.1	2.7	1.8
Se	4	2.2	1.4
Sm	2.6	1.2	0.61
Sn	2	3.8	1.9
Sr	130	35	260
Ta	0.33	0.15	0.15
Tb	0.34	0.22	0.21
Th	4.5	2.1	2.3
Tl		0.66	
U	1.5	1.5	1.2
V	38	32	14

(Continued)

TABLE 3.2 (*Continued*)
Mean Concentration (ppm) of Trace Elements in U.S. Coals

Element	Eastern U.S.	Illinois Basin	Western U.S.
W	0.69	0.82	0.75
Yb	0.83	0.56	0.38
Zn	25	250	7
Zr	45	47	33
No. of samples	23	114	28

Source: Gluskoter et al., 1977.

TABLE 3.3
Global Coal Production and Use in 2010

Type of Coal/Use	M tons	%
Coking coal	891	12.3
Hard coal for steam	5294	73.3
Brown coal/lignite	1042	14.4
Total	7227	100.0

Source: IEA, 2011.

TABLE 3.4
Top Ten Hard Coal Producing Countries in 2010

PR China	3162 Mt	Russia	248 Mt
USA	932 Mt	Indonesia	173 Mt
India	538 Mt	Kazakhstan	105 Mt
Australia	353 Mt	Poland	77 Mt
South Africa	255 Mt	Colombia	74 Mt

Source: World Coal Association, 2011.

steel production (Table 3.3). Over 80% of the hard coal production came from the top 10 producing countries (Table 3.4), while 77% of the worldwide production of brown coal/lignite came from the top 10 producers (Table 3.5).

With an understanding of where coal is mined throughout the world, it is not surprising that those countries that are major coal producers are also major coal users. Table 3.6 summarizes the countries with significant dependence on coal for electrical generation. Each of these countries is also a major producer, which simply illustrates that domestic coal resources are used for electrical generation. This is

TABLE 3.5
Top Ten Brown Coal/Lignite Producing Countries in 2010

Germany	169 Mt	U.S.	65 Mt
Indonesia	163 Mt	Greece	56 Mt
Russia	76 Mt	Poland	56 Mt
Turkey	69 Mt	Czech Republic	44 Mt
Australia	67 Mt	Serbia	37 Mt

Source: World Coal Association, 2011.

TABLE 3.6
Countries with Major Dependence on Coal for Electricity Generation (2008, 2009)

South Africa	93%	Israel	63%
Poland	90%	Czech Republic	56%
PR China	79%	Morocco	55%
Australia	76%	Greece	55%
Kazakhstan	70%	U.S.	45%
India	69%	Germany	44%

Source: World Coal Association, 2011.

particularly true for low rank coals, where the high moisture content (i.e., low energy content) precludes export over long distances. The only countries listed in Table 3.6 that are not major coal producers are Israel and Morocco, both of which import hard coal to meet their electrical generation needs.

3.2 PULVERIZED COAL COMBUSTION BY-PRODUCTS

The most common technology used to generate electricity from coal is pulverized coal combustion. During this process, solid by-products are produced as coal is combusted and any ash contained in the feed coal remains, where it is collected and removed. Additional solid by-products are also generated from treatment of the flue gas to remove sulfur or nitrogen species before the cleaned flue gas is discharged. A simplified schematic is shown in Figure 3.1.

The process begins with pulverizing or grinding the feed coal to a fine size, generally 80% finer than 75 μm prior to pneumatic transfer to burners, where the pulverized coal is mixed with air and combusted in the boiler. Carbon and volatile matter in the coal are consumed and intense heat is generated in the boiler, causing the ash particles introduced with the feed coal to soften or melt. As the softened ash is swept upward, away from the hotter zones of the boiler, the ash solidifies, usually as spherically shaped particulates. Some of the softened ash impinges on boiler tubes

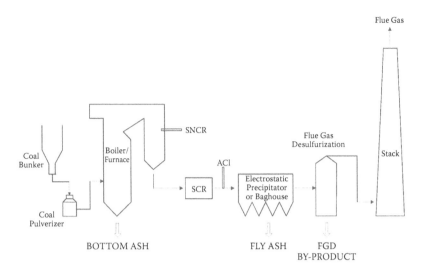

FIGURE 3.1 Simplified schematic of pulverized coal combustion flue gas treatment and by-product production.

and walls, building up and eventually sloughing off as fused masses of agglomerated material, which falls to the bottom of the boiler where it is removed as bottom ash. Most of the ash, generally 80% by mass, exits with flue gas at the top of the boiler. This material is referred to as fly ash.

The flue gas containing the fly ash is sometimes treated for NO_x reduction by a process known as selective non-catalytic reduction (SNCR). SNCR is a post-combustion process that requires a temperature in the range of 900 to 1100°C and usually occurs in the top of the boiler overpass. Ammonia or urea is injected to convert NO_x to N_2 and H_2O (Duo et al., 1992). Another NO_x removal technology that is often used is selective catalytic reduction (SCR), which is applicable at lower temperatures (300 to 400°C). As with SNCR, ammonia compounds are injected to convert NO_x to N_2 and H_2O; however, with SCR, the reaction occurs in the presence of a catalyst to achieve higher efficiency of NO_x conversion. In many cases, sulfur present in the flue gas causes the formation of ammonium sulfate or ammonium bisulfate, which can adversely affect possible uses of the SCR-treated ash.

Another post-combustion process that is often used is activated carbon injection (ACI), a process employed to reduce mercury emissions. Activated carbon is injected into the flue gas to capture speciated mercury by adsorption. The introduction of activated carbon to fly ash, even in small amounts, has an adverse effect on conventional uses for fly ash, which will be described later in this chapter.

The most common industrial method for removing fly ash from flue gas is with electrostatic precipitators (ESPs). Bag houses are sometimes used for smaller volume applications, but ESPs are by far the most widely used. As particulate-laden flue gas enters the ESPs, negatively charged fly ash particulates are attracted to positively charged plates, effectively removing fly ash from the flue gas stream. As ash particles build up on the charged plates, mechanical or pneumatic devices (rappers) periodically strike the plates, dislodging the particles, which flow down the plates into collection hoppers for eventual removal. Most utility boilers utilize several stages of ESPs in order to remove all of the particulate matter from the flue gas before it is emitted.

In many cases, the particulate-free flue gas is finally treated for reduction or removal of any SO_x that is present with some form of flue gas desulfurization (FGD). There are varieties of FGD technologies that are practiced commercially, but they can be classified into two basic approaches—wet and dry. Wet FGD incorporates spraying flue gas with an alkaline sorbent to produce a sulfated solid by-product, while dry FGD uses a dry or slurry-based alkaline sorbent. In most cases, wet FGD is used on flue gas downstream of the ESPs after particulate removal has occurred and the sulfated reaction product is the only by-product from the process. With dry FGD (dry injection or spray drying), SO_x is reacted with the sorbent and reaction products are then removed with particulate control. Since particulate control is necessary, dry FGD is sometimes used on flue gas containing fly ash, but can also be used on flue gas exiting the ESPs.

3.2.1   FLY ASH AND BOTTOM ASH

When pulverized coal is injected into a boiler, combustion occurs in a series of stages that occur over a very short period of time, typically lasting only a few seconds. The first stage is the loss of any moisture associated with the coal upon entry into the hot combustion chamber, followed by combustion of any volatile matter contained in the coal particles. As the volatile matter evolves from the coal matrix, a highly developed surface area remains, consisting primarily of fixed carbon and ash. As the fixed carbon oxidizes, high temperatures result and ash droplets are formed from the softened mineral matter. With the turbulence in the combustion chamber, ash droplets are swept upward into cooler zones where they solidify into particulates that are generally spherically shaped. The spherically shaped particles, known as fly ash, are swept out of the boiler with the flue gas and eventually separated from the flue gas by mechanical devices, such as cyclones, ESPs, or baghouses.

Some of the softened ash droplets formed during combustion impinge onto boiler tubes and form layers of accumulation that cool into hardened masses, which eventually dislodge. If the mass of these agglomerations is too large to be entrained in the flue gas exiting the boiler, it falls to the bottom of the boiler where it is removed as bottom ash. Once removed from the boiler, the bottom ash is either quenched in a water bath or air-cooled and ground to reduce the top size before it is eventually removed by pumping or mechanical conveyance. The proportion of bottom ash is generally 20% of the total ash produced during combustion while fly ash accounts for 80%.

3.2.1.1   Fly Ash Mineralogy and Chemistry

Fly ash is generally regarded as a mixture of three different components in varying concentrations, namely crystalline minerals (quartz, mullite, and spinel), un-burnt carbon, and non-crystalline or amorphous aluminosilicate glass (Ward and French, 2005). The most dominant phase present in fly ash is usually amorphous alumino-silicate glass, thought to be derived from the thermal decomposition of precursor clay minerals (Deer et al., 1992; Araujo et al., 2004). Crystalline phases are typically present in minor concentrations and are generally heterogeneous (Hower et al., 1999a; Nugteren, 2007). Un-burnt carbon has been petrographically characterized to contain three distinct forms: (1) inertinite particles entrained from the boiler prior to combustion or melting, (2) isotropic coke, and (3) anisotropic coke. Both isotropic and anisotropic cokes are extensively reacted particles that appear to have passed through a molten stage (Hower et al., 1995; Hower et al., 1999b).

The chemical composition of fly ash is dependent upon the chemical composition of the inorganic material present in coal from which the fly ash is derived. Studies of trace element partitioning have established a relationship between the chemistry of the feed coal and the fly ash that is produced (Meij, 1994; Mastalerz et al., 2004). While trace element concentrations are important, from a utilization perspective, the bulk chemistry of fly ash primarily dictates utilization. In the United States, the most widely cited chemical specification for utilizing fly ash is the American Society for Testing and Materials (ASTM) designation C618-08 (ASTM, 2008a), which is shown in Table 3.7. With this classification system, Class N refers to raw, calcined, or natural pozzolans such as volcanic ash, opaline cherts, and calcined clays or shales. Class F is fly ash normally produced from burning bituminous or anthracite coal, while Class C is fly ash normally produced from burning lignite or subbituminous coal. The primary chemical difference between Class F and Class C fly ash is the sum of the SiO_2, Al_2O_3, and Fe_2O_3. The lower sum of these chemical constituents allowed for Class C fly ash is typically due to the presence of CaO, hence the term "high calcium fly ash." Generally, Class C fly ash contains 20 to 30% calcium, while the calcium content of Class F fly ash is <5%. Other countries use different chemical standards; however, most generally comply with the classifications outlined in ASTM C618. EU chemical standards for fly ash use in Europe are prescribed by BS EN 450, a specification that was recently rewritten to accommodate increased use of biomass as fuels. New requirements specify that the maximum allowable amount of ash derived from non-coal fuels is 30%, along with specific limits on the amounts of P_2O_5, CaO, and Na_2O in some solid bio fuels (Fuerborn, 2011).

While chemical composition is an important consideration in the utilization of fly ash, perhaps the most important property is pozzolanicity. A pozzolan is a siliceous, or siliceous and aluminous, material that in itself possesses little or no cementitious value, but will, in divided form and in the presence of water, combine with

TABLE 3.7
Chemical Requirements for Pozzolan Use in Concrete (ASTM C618-08)

Property	Class N	Class F	Class C
$SiO_2 + Al_2O_3 + Fe_2O_3$, min%	70.0	70.0	50.0
SO_3, max%	4.0	5.0	5.0
Moisture content, max%	3.0	3.0	3.0
Loss on ignition, max%	10.0	6.0	6.0

Source: ASTM, 2008a.

lime to form cementitious compounds (ASTM, 2008). This property is what makes pozzolans, such as fly ash, useful in Portland cement concrete. Portland cement comprises several phases, one of which is tricalcium silicate, sometimes referred to as alite (Taylor, 1990). When alite hydrates, it forms a cementitious calcium aluminosilicate hydrate gel, which acts as the binder in concrete, and produces free lime as a by-product as shown in the following reaction:

$$2[3CaO \cdot SiO_2] + 7H_2O \rightarrow 3CaO \cdot 2SiO_2 \cdot 4H_2O + 3Ca(OH)_2$$

The role of a pozzolan is to react with the free lime to produce additional cementitious calcium aluminosilicate hydrate gels, which act as additional binder. Because of this pozzolanic reaction, fly ash can be used as a substitute for a portion of the Portland cement in concrete, contributing to long-term strength development and other durability benefits.

3.2.1.2 Fly Ash Production and Utilization

While the amount of coal that is utilized throughout the world is annually documented by numerous organizations and government agencies, the amount of coal ash produced and utilized is not. Some countries do document ash production and utilization, but no single source compiles this data. A comprehensive estimate of worldwide coal ash production and utilization was published by Manz in 1997 and reported that in 1992, ~492 million tonnes were produced and ~153 million tonnes, or 33.3%, were utilized. A more recent estimate was made using data available from various national coal ash organizations such as the American Coal Ash Association (ACAA) and the Ash Development Association of Australia (ADAA), while additional data was obtained from journal articles. For countries that did not have coal ash production data readily available, coal utilization and quality data were used where available. By doing so, it is estimated that annual coal ash production increased to ~720 million tonnes in 2010, of which approximately 285 million tonnes, or 40%, were utilized. Major coal ash producing countries are summarized in Table 3.8. These countries accounted for more than 95% of the estimated coal ash produced. The amount of coal ash produced is not only dependent on the amount of coal used, but is also a function of coal quality. This is particularly true for countries such as India and South Africa, which utilize domestic coal resources that have very high ash content. For example, typical run-of-mine coal in India has an ash content of 40 to 50% (Krishna, 1980; Visuvasam et al., 2005). It is generally true that 80% of the coal ash produced by a utility is fly ash, while 20% is bottom ash. The uses of these two products can be quite varied and will be described separately.

3.2.1.2.1 Cement, Concrete, and Grout

Cementitious applications are the major use for fly ash. Specific applications include cement, concrete, and grout, and concrete is by far the largest. These applications exploit the pozzolanic properties of fly ash that can improve concrete strength and durability. In addition, massive pours, such as in bridge foundations, roads, and dams, incorporate fly ash in the concrete mix design to control heat of hydration during curing. In most applications, fly ash is used as a partial replacement (20%) for

TABLE 3.8
Major Coal Ash Producing Countries

Country	Coal Ash Production (M tonnes)	Coal Ash Utilization (M tonnes)	Coal Ash Utilization (%)	Year Reported	Ref.
China	320.0	190.0	59.4	2010	Cao et al., 2008
India	130.0	60.0	46.2	2007	Singh, 2010
United States	79.7	31.5	39.5	2010	ACAA, 2010
EU15	48.9	23.8	48.8	2010	Fuerborn, 2010
South Africa	29.0	1.1	3.8	2010	Eskom, 2010
Russia	27.8	5.0	18.0	2005	Putilov and Putilova, 2005
Ukraine	20.3	n.a.	n.a.	2009	Coalukraine.com, 2009
Japan	10.0	9.7	97.2	2006	JCOAL, 2006
Australia	14.1	5.8	41.3	2010	ADAA, 2010
Turkey	15.0	0.2	1.0	2006	Sezer et al., 2006
South Korea	5.4	3.2	58.0	2010	Kim et al., 2011
Canada	5.8	1.8	31.0	2006	CIRCA, 2006

Portland cement, although higher substitution rates can be used, depending on the application. Specific benefits derived from using fly ash include improved workability, decreased water demand, increased ultimate strength, reduced permeability, and improved durability (ACAA, 2003). In addition, cost savings are provided as fly ash is generally less expensive than Portland cement.

3.2.1.2.2 Raw Feed for Portland Cement Clinker

Portland cement is the basic ingredient in concrete and is produced by heating a combination of oxides of calcium, silicon, aluminium, and iron to produce cement that meets specific chemical and physical specifications. The raw ingredients are mixed together and heated in a rotary kiln to 1430 to 1650°C to induce a series of chemical reactions that cause the raw materials to form cement clinker, a pelletized product that is cooled and ground to a powder consistency that is known as Portland cement (Portland Cement Association, 2012). In most cases, raw ingredients include combinations of limestone, shells or chalk, shale, clay, sand, and iron ore. The raw ingredients are selected based upon availability to obtain the desired amounts of Ca, Si, Al, and Fe to produce clinker with the desired composition. Since coal ash contains these same components in varying concentrations, it is often incorporated into the kiln feed, particularly when location and availability are advantageous.

3.2.1.2.3 Blended Cement

Blended cements are produced by inter-grinding or blending two or more types of fine materials. The most common materials blended are Portland cement, slag, fly ash, silica fume, calcined clay, and hydrated lime. In the United States, specifications are outlined in ASTM C595 (ASTM, 2008b), which recognizes three different classes of blended cements which can contain as much as 40% pozzolan, such as fly ash.

3.2.1.2.4   Flowable Fill

Flowable fill is also known as controlled low-strength material (CLSM) and controlled density fill (CDF), which is a mixture of cement, fly ash, sand, and water. Sand is the principal ingredient of a mixture that is designed as a low-strength, flow-able material that will achieve 100% consolidation without vibration or tamping (Swan et al., 2007). Common applications are backfilling sewer and utility trenches, and structural fill for pipe bedding. Some of the advantages of flowable fill are that it is easily placed, will not settle, does not require compaction around pipes and tanks, and can be easily excavated with conventional digging equipment. While flowable fills achieve a typical unconfined compressive strength of 8 MPa, most applications require strengths of less than 2 MPa (Rajendran, 1994). Other uses include most types of construction activities that require backfilling. One particular advantage is that many fly ashes that do not meet ASTM C618 specifications because of high carbon content can readily be used in flowable fills.

3.2.1.2.5   Structural Fills/Embankments

Structural fill is a designed fill material that is placed to achieve specific engineering properties. Specifications are similar to those for engineered soil fills and generally focus on moisture and compaction to achieve the desired strength and compressibility required for the application. For U.S. road embankments, typical requirements are compaction to 95 to 100% of maximum dry density as determined by ASTM D698 (ASTM, 2008d). A general overview of considerations for placement is provided by the American Coal Ash Association (2003) while the equipment required to achieve desired compaction is described in AASHTO Method T 99 (2011).

3.2.1.2.6   Road Base/Subbase

Base courses for construction applications such as roads and foundations consist of proportioned mixes of fly ash, bottom ash, other aggregate, and in some cases, where stabilization is desired, an activator such as lime, cement, or Class C ash. When the mixture is properly placed and compacted, the result is a strong and durable base course. The use of fly ash to stabilize base courses is sometimes referred to as a pozzolanic-stabilized mixture (PSM) and can incorporate several material combinations. Class C ash can be used alone while Class F ash requires the addition of Portland cement or cement kiln dust. Typical proportions for Class F ash are 2 to 8% lime or 0.5 to 1.5% Portland cement blended with 10 to 15% fly ash (ACAA, 2003).

3.2.1.2.7   Soil Modification/Stabilization

Class C fly ash and Class F-lime blends can be used to alter soil properties in order to improve soil properties such as density, moisture content, plasticity, and strength. If such changes are temporary in order to expedite construction, the practice is termed soil modification, while more permanent changes in soil properties are considered soil stabilization. While these practices are used for numerous geotechnical reasons, some of the more common are enhancing soil strength, controlling swelling of expansive soils, and reducing moisture content to facilitate compaction. As with road base and sub-base applications, Class C ash can be used alone, while Class F ash requires a cementitious additive such as lime or cement. The self-cementing behavior of fly ash is determined by ASTM D5239 (ASTM, 2008c), which describes a standard method for determining compressive strength of cubes made with fly ash and water, which are tested after curing for 7 days. The self-cementing characteristics are classified into the following categories:

Very self-cementing   >3400 kPa (500 psi)
Moderately self-cementing   700–3400 kPa (100–500 psi)
Non self-cementing   <700 kPa (100 psi)

As far as stabilization of soils is concerned, fly ash is often used to stabilize subgrades or bases, reduce lateral pressures of backfills, and improve slope stability of embankments. The primary reason fly ash is used in these applications is to improve the compressive and shear strengths of soils, typically by the addition of 15 to 46 cm (6 to 18 in.) of ash (ACAA, 2003).

3.2.1.2.8   Waste Stabilization/Solidification

Waste stabilization and solidification use applications follow similar guidelines to those pertaining to soil stabilization and solidification. Some examples of wastes where this application has been successfully demonstrated include hazardous waste (Parsa et al., 1996), foundry waste (Reddi et al., 1996), and wastewater treatment sludge (Burns and Gremminger, 1994). A primary advantage of using ash in these applications is that self-cementing reactions of Class C ash limit the mobility of water containing undesirable elements by utilizing the water in cementitious reactions.

Geopolymers, or alkali-activated cements, have received considerable attention for these applications. Geopolymers utilize cementitious reactions of pozzolans such as fly ash to produce a dense concrete matrix without Portland cement. While similar to cement encapsulation for immobilizing metal ions, geopolymerization can offer improved strength and durability while reducing permeability (Davidovits and Comri, 1988; Jaarsveld et al., 1998; Davidovits, 1994). More recently, geopolymerization of fly ash has been shown to be effective for stabilizing hazardous elements in electric arc furnace dust (Luna et al., 2009), as well as Cr, Cd, and Pb contaminated water (Zhang et al., 2008).

3.2.1.2.9   Blasting Grit/Roofing Granules

These uses almost exclusively use cyclone boiler ash rather than pulverized combustion ash. Cyclone boilers differ in that they burn poorer quality fuels at higher temperatures and produce a slag-like by-product that is removed from the boiler in a molten state and rapidly quenched. The quenched slag is typically dense, fine-grained, and hard, making it suitable for uses such as blasting grit and roofing granules.

A key advantage of boiler slag for use as blasting grit is that it contains essentially no crystalline silica, thus minimizing risks of respiratory diseases such as silicosis. Other advantages are that boiler slag has very low porosity and is generally rounded, which minimizes abrasion. Blasting grit specifications depend on the application, but generally focus on size gradation and hardness. In asphalt shingles, boiler slag is mostly used as headlap, the portion of the shingle's exterior surface that is overlapped by the shingle above it when applied to a roof. It is also used in the visible portion of the shingle because it has excellent weathering characteristics and is opaque to UV penetration, which degrades the asphalt binder (ACAA, 1994).

3.2.1.2.10   Snow and Ice Control

The angular shape of bottom ash makes it particularly useful as anti-skid material on roadways, particularly if the bottom ash is not friable. This application can use both fresh and stored material, which is applied with conventional spreaders. Other specific advantages are low cost, local availability, and, most notably, bottom ash can be stored in outdoor stockpiles for indefinite periods of time without degrading.

3.2.1.2.11   Mining Applications

Coal ash can be beneficially utilized in both surface and underground mining in numerous ways. The most common applications include void filling to control sub-sidence and groundwater movement, neutralizing acidic water, and soil remediation during reclamation (Ward et al., 2006). Fly ash slurry can be injected into abandoned mines to prevent subsidence. Because fly ash particles are spherical, 50 to 60% solids can readily be pumped and can flow significant distances underground. It has been reported that fly ash slurries can travel more than a kilometer from their injection point (Lokhande et al., 2005). In this manner, large areas may be backfilled from a single injection point. By effectively backfilling mined areas in underground mines, surface subsidence effects can be minimized or prevented.

Flowable fill made from coal combustion by-products (CCBs) has been success-fully used to seal stream loss zones in karst topography in abandoned coalmines (Reeves, 2009). Backfilling of coalmines with CCBs has also been shown to reduce acid production in mine openings subject to the passage of water (Lee, Ranalko, and Giacint, 2008). Similar beneficial acid mitigation results were demonstrated using surface grouting rather than backfilling (Guynn et al., 2009). Surface grouting offers the advantages of preserving the flow system within the mine while decreasing the possibility of blowouts that can occur with bulk fills.

3.2.1.2.12   Agriculture

Fly ash has been used for many years as a soil amendment for agricultural purposes. Jala and Goyal (2006) reported that fly ash in conjunction with organic manure and microbial inoculants can enhance biomass production from depleted soils. Application of fly ash can be used to increase the pH of acidic soils (Phung et al., 1979) and improve soil texture (Chang et al., 1977). When mixed with soil, fly ash increased crop yields for alfalfa, barley, Bermuda grass, and white clover by improving the physical and chemical characteristics of the soil (Weisenstein et al., 1989). A number of forestry species have also been found to grow and establish well on fly ash overburdens (Adholeya et al., 1998). Fly ash added to poorly drained sandy soil increased run-off pH and doubled the growth rate of Australian pine (Riekerk, 1984).

Significant research has been focused on utilizing fly ash for wasteland reclamation. It is often used as an alternative to lime to neutralize acidic mine soils (Sing et al., 1982; Haering and Daniels, 1991; Carlson and Adriano, 1993). The addition of fly ash also decreases soil bulk density, increases water-holding capacity, and reduces compaction (Fail and Wochok, 1977; Capp, 1978; Jastrow et al., 1981).

3.2.1.2.13 Aggregate

Bottom ash is frequently used as a substitute or partial replacement for conventional aggregate in concrete and concrete masonry units (CMUs), more commonly known as blocks. Quality considerations include assurance that the aggregate is not friable and will retain structural integrity under compression. Another quality consideration is that pyritic material rejected from coal pulverizers at the power plant be isolated from the bottom ash. Co-mingled pulverizer rejects can result in iron staining and cause pop-outs in finished masonry. Drained bottom ash is screened to meet gradation requirements for the intended applications.

Frequently, bottom ash is porous and has a lower bulk density than conventional aggregates, which makes it particularly useful in producing lightweight concrete and lightweight CMUs. Specifications for lightweight aggregate for CMUs are prescribed by ASTM C331 (ASTM, 2010) and are summarized in Table 3.9. Fine aggregate is the most common size designation for bottom ash. Traditional lightweight aggregates used in this application, such as slate, shale, and clay, are mined materials that require heating in a kiln to generate porosity. It has been estimated that the energy required to produce lightweight aggregates from traditional aggregates is 3 GJ per tonne of aggregate produced (Haseltine, 1975), which generates 0.3 tonnes CO_2 per tonne of aggregate produced (Malhotra, 1986). Lightweight bottom ash offers the advantage of no energy requirements for mining and heating because it is strictly a recycled by-product.

3.2.1.2.14 Miscellaneous Uses

Coal ash can be used in a variety of building applications including landscape furniture, manufactured stone, ceiling tile, carpet backing, flooring tile, and tile underlayment (USEPA, 2008). LifeTime Lumber® is an example of a wood alternative made with up to 60% recycled fly ash (LifeTime Composites, 2012). Because this type of product contains no wood, it is resistant to water absorption, mold, and rot, while

TABLE 3.9

Maximum Bulk Density (Loose) Requirements of Lightweight Aggregates for Concrete Masonry Units

Size Designation	Maximum Dry Loose Bulk Density	
	kg/m³	lb/ft³
Fine aggregate	1120	70
Coarse aggregate	880	55
Combined fine and coarse aggregate	1040	65

offering strength, durability, and flexibility. CERATECH (2012) offers a complete line of cement and rapid hardening concrete repair products produced primarily from fly ash. Other uses include plastic functional filler for injection-molded plastics, prepackaged cement, and concrete for residential applications and as an additive to abrasion-resistant paints.

3.2.1.2.15 Utilization Statistics

According to statistics compiled by the American Coal Ash Association (ACAA, 2010), nearly 80 million tonnes of fly ash, bottom ash, and boiler slag were produced in 2010 and 31.46 million tonnes were utilized, a utilization rate of 35.8% (Table 3.10). Over one-third of the utilization was for cement, concrete, and grout, while the next largest usage category was structural fill (22.7%). A comparison of coal ash, bottom ash, and boiler slag utilization between several different countries is shown in Table 3.11. To simplify comparisons between different countries, data are presented as a percentage of total utilization. Although each country reports different utilization categories, some general similarities and significant differences are evident. For example, cement, concrete, and grout account for approximately 70% of the ash utilization in Australia, Japan, and Canada and much less (only approximately one-third) for the United States and Europe. This may be due to the significantly larger quantities of ash produced in the United States and Europe; therefore, other uses are more developed, such as structural fill, accounting for

TABLE 3.10
U.S. 2010 Fly Ash, Bottom Ash, and Boiler Slag Production and Utilization

Utilization Category	M tonnes Utilized	% of Total Utilized
Cement/concrete/grout	10.55	33.5
Raw clinker feed/blended cement	2.72	8.6
Flowable fill	0.17	0.5
Structural fill	7.15	22.7
Road base/subbase	0.87	2.8
Soil modification	0.86	2.7
Snow and ice control	0.54	1.7
Blasting grit	1.24	3.9
Mining applications	2.66	8.4
Waste stabilization/solidification	2.99	9.5
Agriculture	0.02	0.1
Aggregate	0.53	1.7
Total	31.46	100.0
Total produced, tonnes	79.66	
% Utilized	35.82	

Source: ACAA, 2010.

TABLE 3.11

Comparison of Coal Ash Utilization Practice of Different Countries

Country	Australia	Japan	Canada	EU15	U.S.
Year	2008	2006	2010	2009	2010
Ref.	ADAA (2010)	JCOAL (2006)	CIRCA (2006)	ECOBA (2011)	ACAA (2010)
Cement, concrete, grout	71.0	71.0	67.8	29.3	33.5
Blended cement, raw clinker feed				31.4	8.6
Flowable fill	24.0				0.5
Mining applications	5.0	2.0	7.8		8.4
Road base/subbase			9.6	3.9	2.8
Soil treatment, modification		3.8		0.2	2.7
Soil stabilization		3.0			9.5
Blasting grit, roofing granules				2.8	3.9
Structural fill		5.9		21.3	22.7
Aggregate				5.6	1.7
Other		14.3	14.9	5.5	5.7
Total	100.0	100.0	100.0	100.0	100.0

over 21% in both regions. Flowable fill is widely practiced in Australia, and mining applications are significant in countries where mining is prevalent. Another notable difference is that use of ash in blended cements is common in Europe, but not as widely practiced in other countries.

3.2.2 WET FLUE GAS DESULFURIZATION (FGD)

FGD refers to the technologies employed to remove gaseous sulfur species from flue gas. Various sulfur species, derived from combustion of organic and inorganic sulfur present in coal, contribute to acidic rainfall and need to be removed from flue gas before it is emitted from the power plant in order to meet strict emission standards. FGD technologies were developed during the 1970s and have been widely adopted over the past few decades. Wet FGD refers to technologies where flue gas is treated with a sorbent suspension or liquid with a high liquid-to-gas ratio of sorbent in order to remove SO_x generally after fly ash has been removed. Technologies include spray towers, venturis, plate towers, and mobile packed beds. Most wet FGD systems currently in use are spray towers, where the flue gas flows co-current, counter-current, or cross-current to the absorber liquid. Sorbents commonly used include limestone ($CaCO_3$), lime ($Ca(OH)_2$), and magnesium hydroxide ($Mg(OH)_2$).

3.2.2.1 Wet FGD By-Product Mineralogy and Chemistry

Calcite is a common constituent in limestone, which is widely available, making it a useful sorbent for wet FGD. Limestone is ground to a fine consistency and added into the absorber as slurry. Calcite reacts with SO_2 to form either gypsum or hannebachite, depending upon the amount of oxygen present. In an oxygen-deficient environment, hannebachite is formed as follows:

$$CaCO_3 \text{ (calcite)} + SO_2 + \tfrac{1}{2}H_2O \rightarrow CaSO_3 \cdot \tfrac{1}{2}H_2O \text{ (hannebachite)} + CO_2$$

If sufficient oxygen is present, then the following reaction occurs:

$$CaCO_3 \text{ (calcite)} + SO_2 + \tfrac{1}{2}O_2 + 2H_2O \rightarrow CaSO_4 \cdot 2H_2O \text{ (gypsum)} + CO_2$$

Since gypsum is potentially useful as a by-product and hannebachite is not, many wet FGD systems using limestone as a sorbent incorporate forced oxidation to ensure that the sorbent reaction produces gypsum. If lime is used as the sorbent, the reaction produces calcium sulfite:

$$Ca(OH)_2 \text{ (lime)} + SO_2 \rightarrow CaSO_3 \text{ (calcium sulfite)} + H_2O$$

If magnesium hydroxide is used, magnesium sulfite is formed:

$$Mg(OH)_2 \text{ (magnesium hydroxide)} + SO_2 \rightarrow MgSO_3 \text{ (magnesium sulfite)} + H_2O$$

Of the sorbent reaction products produced with wet FGD, gypsum is the only one that is utilized in significant quantities. If the other by-products are produced, they are generally dewatered and impounded in landfills.

3.2.2.2   Wet FGD By-Product Utilization

Use of wet FGD by-products almost exclusively pertains to gypsum, the product of forced oxidation scrubber systems. This synthetic gypsum, often called "syngyp" is mineralogically and chemically the same as natural gypsum, so it is used in many of the applications where natural gypsum is used.

3.2.2.2.1   Cement

Ordinary Portland cement (OPC) is the most common cement product used for most purposes. Gypsum is sometimes added to OPC to regulate set time (Murakami, 1968). Several other types of cement are also produced for a variety of specialty applications. Supersulfated cements are produced by inter-grinding gypsum (~15%) with OPC clinker and blast furnace slag (Lea, 1970). Supersulfated cements are often used when sulfate resistance is an important durability consideration, such as marine construction (Uomoto and Kobayashi, 1983).

Calcium sulfoaluminate cements (CSAs) are ultra-high early strength cements that also use significant quantities of gypsum. These cements require less energy to produce than OPC and their development has been pioneered in China, where several million tonnes per year are produced (Bye, 1999; Zhang et al., 1999). Lower kiln temperature is required for their production, and hence lower energy consumption. Additionally, less limestone is required as a raw ingredient, and thus less CO_2 is produced during the manufacturing process.

When gypsum is heated to ~150°C, plaster of Paris or calcium sulfate hemihydrate ($CaSO_4 \cdot \tfrac{1}{2}H_2O$) is formed (Appleyard, 1975). If heated to ~200°C, anhydrite is formed (Deer et al., 1992). When mixed with water, both plaster of Paris and anhydrite will hydrate back into gypsum, which hardens upon drying. While plasters can become very hard, they have little durability in the presence of water and their use is limited to water-free environments. Most commonly, gypsum plasters are used in interior finishing because they can be sanded to a smooth finish when dry. When used in exterior coatings, they are usually mixed with OPC to provide weathering resistance. Gypsum plasters can also be applied as a sprayed interior layer in building to provide insulation, fireproofing, and acoustic dampening. Plaster is often used in artistic applications because it can be poured into molds to create hardened objects that can be sanded and painted. It has also been used in medicine for many years to create a hardened cast to immobilize broken bones. Bandages are soaked in plaster and wrapped around the injured limb to create the hardened cast.

3.2.2.2.2   Stabilized Base/Embankments
Wet scrubber by-products have been successfully used for road base construction at numerous locations (Smith, 1985, 1989; Amaya et al., 1997; Prusinski et al., 1995). Dewatered gypsum and calcium sulfite, as well as a mixture of both, can be used in this application provided they are fixated with quicklime, OPC, or fly ash (Class F or Class C). To achieve compressive strength or durability specifications, additional fixating agents can also be used (USFHWA, 1997). Fixated FGD scrubber material has also been used for embankment construction (Brendel and Glogowski, 1989). Reclaimed FGD material was used in conjunction with a fly ash as the fixating agents.

3.2.2.2.3   Mine Reclamation
Wet FGD materials have been used in numerous beneficial ways in mine reclamation. The approach to these uses is similar to the approach used in stabilized bases and embankments and incorporates a fixing agent such as lime or fly ash to solidify the FGD into a strong, low permeability product. Some of the uses are intermediate and final cover for coal waste disposal areas, caps for unreclaimed abandoned coal refuse piles, pond liners, grout mix barriers to abate acid mine drainage, and structural fills for abandoned highwalls (Wolfe et al., 2009). Environmental impact studies of these types of applications demonstrate no significant impact on surrounding waters (Pasini, 2009).

Of all the mine reclamation benefits demonstrated, perhaps the most widely investigated has been acid mine drainage (AMD) mitigation (Lamminen et al., 1999; Lyons and Petzrich, 1995; Meiers et al., 1995). Injection of fixated FGD materials has been shown to reduce water flow from abandoned mines by creating a physical barrier to flow, while providing alkalinity to neutralize the acidity of AMD.

FGD materials have also been shown to promote good long-term vegetative growth in mine reclamation soils, while significantly improving water quality (Chen et al., 2009). Growth benefits were attributed to high concentrations of essential macronutrients calcium, magnesium, and sulfur in the FGD, as well as the essential micronutrients iron and boron. Drainage water quality was also improved over prereclamation quality; electrical conductivity and concentrations of magnesium, sulfur, and boron in drainage water increased after FGD application and the increase in pH when alkaline materials were applied precipitated iron, aluminium, and other metals.

3.2.2.2.4 Gypsum Panel Products

Gypsum panel products, otherwise known as wallboard, comprise the largest utilization market for FGD gypsum. Wallboard is essentially a flat, thin slab of stucco (gypsum calcined to calcium sulfate hemi-hydrate, $CaSO_4 \cdot \frac{1}{2}H_2O$) cast as thickened water-based slurry, sandwiched between paper bonded to both sides and edges (Appleyard, 1975). Accelerators, foaming agents, and fibers can also be added into the slurry, but the primary raw ingredient is gypsum. In the manufacturing process, dry ingredients are mixed with water and the resulting slurry is spread onto a moving strip of paper. An additional strip of paper is added onto the top of the moving mass and squeezed to the desired thickness with rollers. The sandwiched board travels along a flat conveyor that is several hundred meters long to provide sufficient time (5 to 6 minutes) for the stucco to set. The continuous board is then cut into shorter lengths, and residual moisture is removed by heating with hot air.

Using FGD gypsum as replacement for natural gypsum offers numerous energy and environmental benefits. Sustainable wallboard manufacturing with FGD gypsum avoids the impacts of mining virgin ore and landfilling scrubber by-products, while annually saving 1200 million MJ of energy, 18 billion liters of water, and reducing greenhouse gas (GHG) emissions by 83,000 Mg CO_2e. Annual cost savings are estimated to be $49M to $63M (Lee et al., 2011).

3.2.2.2.5 Agriculture

FGD gypsum offers numerous agricultural benefits by improving soil properties, providing nutrients, and even removing phytotoxic elements. Over the past 30 years, the amount of sulfur deposited by rainfall in the United States has been reduced by about half. This decrease coupled with other decreases in sulfur inputs to soil due to the use of concentrated fertilizers containing little or no sulfur and intensive crop yields have resulted in more sulfur removal from soil, leading to deficiencies in crops. Because gypsum solubilizes slowly, it can continuously release sulfur into soils where it is applied, enhancing production of crops such as corn, soybean, canola, and alfalfa (Chen and Dick, 2011). Since calcium uptake through the plant is a very slow process, calcium must be continually available to the root system. Thus, gypsum can be beneficial to crop quality (Scott et al., 1993; Shear, 1979).

Gypsum can also be used to improve soil quality, particularly in clay soils. Gypsum reduces the dispersion of clays that lead to crust formation and slows the rate of surface drying, both of which have been shown to improve seedling emergence (Norton et al., 1993). Root growth has also been shown to be improved in calcium-deficient soils as calcium from gypsum migrates into the subsoil (Farina and Channon, 1988; Toma et al., 1999).

A common use for gypsum is to reclaim sodic soils, as Ca^{2+} exchanges with Na^+ and Mg^{2+}, leading to flocculation of soil particles, which improves drainage properties (Chen and Dick, 2011). FGD gypsum has also been used to ameliorate the phytotoxic conditions arising from excess soluble aluminium in acid soils by reacting with Al^{3+}, thus removing it from the soil solution and greatly reducing its toxic effects (Shainberg et al., 1989; Smyth and Cravo, 1992).

3.2.3 DRY FGD

Dry scrubbing is a process where an atomized alkaline sorbent (lime or soda ash) slurry is mixed concurrently with flue gas and reacted in an absorber, thus the name spray dryer absorber (SDA). Sorbent droplets react with SO_2 in the flue gas and the desulfurized gas, along with reaction products and unreacted sorbent, exit the absorber where they are collected in a baghouse. Dry scrubbers can be used on flue gas prior to fly ash collection, so the existing particulate collection system can often be used to collect fly ash and absorber products. A specific advantage is that dry scrubbers are less expensive to construct, operate, and maintain than wet scrubbers, making them particularly suitable for smaller power plants. Additionally, the by-product is dry, significantly reducing by-product handling requirements. Sorbent requirements to achieve comparable SO_2 reduction are lower than for wet FGD, but dry scrubbing requires a more reactive sorbent such as lime or soda ash, which is more costly (Gaikwad, 2002).

An emerging dry FGD technology is circulating fluidized bed-flue gas desulfurization (CFB-FGD), which has recently been installed at several sites in the United States and has been applied in China at power plants up to 660 MW (Jiang et al., 2011). In addition to water and energy conservation and smaller footprint, this process has the capability of multi-component control (Connell, 2009).

3.2.3.1 Dry FGD By-Product Mineralogy and Chemistry

The chemistry and mineralogy of dry scrubber by-products are widely dependent on a number of factors including the type of alkaline sorbent used, boiler fuel characteristics, and, most notably, whether it is collected as a sole product or in conjunction with fly ash. Examples of SDA composition are presented in Table 3.12, which shows enrichment in Ca species present, most notably $CaSO_3$, the reaction product of the sorbent and SO_2. In general, when SDA ash is collected as a sole product, it is finer grained than fly ash and predominantly crystalline. When combined with fly ash, it is a combination of glassy fly ash particles coated by and intermixed with fine crystals of calcium/sulfur reaction products. Generally speaking, most dry FGD by-products contain little or no fly ash and the major component is calcium sulfate hemi-hydrate ($CaSO_3 \cdot \frac{1}{2}H_2O$) (Heebink et al., 2007a).

3.2.3.2 Dry FGD By-Product Utilization

3.2.3.2.1 Structural Fill, Embankments, Road Base

Blends of SDA ash, fly ash, and OPC (30 to 70% SDA ash, 30 to 70% fly ash, and 0 to 8% OPC) are sometimes referred to as stabilisate products (Heebink et al., 2007b) and have been used as capping material for metal ore mining residues and as a vertical seal for preventing groundwater contamination. The stabilized product has moderate compressive strength (5 to 35 MPa, 725 to 4350 psi) and low permeability (10^{-10} to 10^{-12} cm/sec), making it suitable for these applications. Stabilisate products

TABLE 3.12

Composition of SDA By-Products

Analysis	SDA Ash (no fly ash) (%)	SDA-Fly Ash Blend (%)
CaO (total)	44.4	27.8
Available CaO	1.7	4.3
MgO	0.3	0.7
SiO_2	1.1	20.9
Al_2O_3	0.2	10.5
Fe_2O_3	0.2	6.3
$CaSO_4$ (anhydrite)	5.61	3.91
$CaSO_4$	66.75	33.19
$CaCO_3$	16.5	11.4
Bulk density, kg/m^3 (lb/ft^3)	560 (35)	688 (43)

Source: Beeghly and Schrock, 2009.

have been extensively used in Europe in base course applications, as well as land reclamation and landscaping (Bengtsson, 2001).

Suitability for structural fill applications has also been evaluated (Dawson et al., 1987). Reactive SDA ash achieved compressive strengths in excess of 13.8 MPa (2000 psi) while nonreactive material strengths were much lower. Swelling is of primary concern when placing SDA ash. It is important that reactive components, particularly free lime and anhydrite, be properly hydrated before or during initial placement. Formation of secondary minerals such as etteringite can cause excessive swelling (Dick et al., 1999). A criterion for compaction after hydration in road base and embankment applications is to a minimum of 70% of Proctor density (Webster, 1982).

3.2.3.2.2 Fixating Agent for Waste

Alkaline FGD materials have been demonstrated to be useful in stabilizing metals in hazardous waste, particularly those that have reduced solubility at high pH (i.e., Cd, Fe, Mn, Zn, Cu, and Co) (Brendel et al., 1997). Waste stabilization practice is similar to road base stabilization; however, there is higher moisture content in waste sludge. Adding dry SDA materials thickens the sludge and limits mobility. Leaching studies have shown that SDA ash could potentially be used as an additive in the co-disposal of chemical wastes containing elements whose leachability decreases with increasing pH (Farber et al., 1983).

3.2.3.2.3 Binders

SDA ash has been used as a component in binders for interior plasters (Koslowski and Roggendorf, 1996). The specific component required for this application is calcium sulfite. A different interior binder is prepared by mixing SDA ash with fly ash, followed by pre-calcining, oxidation, and calcining, followed by grinding with other additives to produce a binder used in flooring and insulating building materials (Kolar, 1995).

3.2.3.2.4 Cement Manufacture

SDA material has been used commercially in Germany to manufacture cement. SDA ash is treated in a fluidized bed to produce pelletized anhydrite, which can be used as a substitute for natural anhydrite (vom Berg et al., 1993). Both the Fläkt-Dorr-Oliver process and the Vereinigte Aluminium-Werke AG, Lünen (VAW) process produce anhydrite from SDA material (Kolar, 1995). Another, in the Müller-Kühne process, uses dry FGD material to manufacture cement, rather than anhydrite alone (Bengtsson, 2001). SDA material can also be used as a substitute for fly ash and gypsum in cement (Bloss, 1984). In this approach, calcium sulfite hemihydrate is used in place of gypsum to control set time.

3.2.3.2.5 Cement and Concrete

SDA has been shown to be suitable for utilization in structural concrete, providing increased compressive and bond strengths as well as improvements in freeze–thaw and corrosion resistance. The material exhibited performance properties similar to Class C fly ash and OPC (Namagga, 2010). SDA ash and crumb rubber were evaluated as structural materials in concrete acoustic barriers (Heyliger et al., 2011). Results showed that SDA ash improved long-term strength while reducing material costs. Blended cements using as much as 40 to 80% Class C ash and SDA ash, along with sodium sulfate anhydrite as a chemical activator, showed higher early strength development than OPC as well as improved resistance to both sulfate attack and alkali-silica reactions (Wu and Naik, 2003). High content SDA concrete (50%) used in concrete door portal frames has also been shown to perform as well as conventional concrete in structural members for buildings in high seismic areas (van de Lindt and Rechan, 2011).

3.2.3.2.6 Utilization Statistics

Utilization of FGD products has increased significantly as adoption of flue gas desulfurization technologies has become more widespread throughout the world. A significant driver for increased utilization has been the need to develop sustainable utilization practices as the quantity of these by-products significantly increases. Research and development activities have greatly improved our understanding of the composition and behavior of these materials. In the United States, the amount of FGD by-products increased by 7.3% between 2006 and 2010 while coal consumption actually decreased by 5.5%. Decreased coal consumption during this time is primarily attributed to decreased electricity demand caused by a slow-down in economic activity (EIA, 2011). Between 2006 and 2010, the amount of wet scrubber by-product decreased by 47% while the amount of gypsum generated increased by 82% as additional FGD forced oxidation systems were added by both new construction and conversion. The amount of dry FGD by-products produced remained essentially the same (ACAA, 2010). It is clear that FGD is becoming more widely employed in U.S. power plants to meet air emission regulations.

TABLE 3.13

2010 U.S. FGD Byproduct Production and Utilization, Tonnes

Utilized for	FGD Gypsum	Wet Scrubber	Dry Scrubber	Total
Concrete/concrete products/grout	19,105	—	15,294	34,398
Blended cement/raw clinker feed	1,030,545	—	—	1,030,545
Flowable fill	—	—	12,707	12,707
Structural fills/embankments	412,532	385,435	325,010	1,122,976
Road base/subbase	—	2740	—	2740
Soil stabilization	—	—	17,420	17,420
Mining applications	758,500	169,417	102,012	1,029,929
Gypsum panel products	6,955,134	—	—	6,955,134
Waste stabilization	—	—	35,661	35,661
Agriculture	437,403	—	—	437,403
Other	112,170	9078	22,153	143,401
Total utilized	9,725,387	566,670	530,257	10,822,313
	—	—	—	—
Total produced	19,971,600	7,871,365	1,276,323	29,119,288
% Utilized	48.7	7.2	41.5	37.2

Source: ACAA, 2010.

As shown in Table 3.13, of the 29.1 million tonnes of FGD by-products produced, nearly half (48.7%) was classified as FGD gypsum, while 41.5% was dry FGD, and the remaining 7.2% was wet FGD other than gypsum (ACAA, 2010). The single largest use for FGD gypsum in the United States was wallboard products, which accounted for 48.7% of the 20 million tonnes produced. This market has shown continued growth over the past two decades, as shown in Figure 3.2. Major use (over 60%) for other wet and dry FGD by-products is in structural fill and embankments (ACAA, 2010).

In Europe, 8.8 million tonnes of FGD gypsum was produced in 2008 and 63% was used in wallboard. Other applications included the production of gypsum blocks, plasters, and self-leveling floor screeds. Only 0.2 million tonnes of SDA ash was used, primarily in structural fill (Feuerborn, 2010).

3.3   FLUIDIZED BED COMBUSTION (FBC)

Fluidized bed combustion (FBC) is a technology whereby solid fuel is fluidized by air in a combustion chamber, producing turbulent mixing of fuel and gases. The fuel is typically coal, biomass, or a mixture of both. A significant difference from pulverized coal combustion is the velocity of the air used, generally high enough to fluidize the fuel bed. FBC technologies achieve very low SO_x emission levels by adding a sorbent, such as limestone, into the fluidization vessel so that sulfur capture occurs during combustion, thus eliminating the need for downstream flue gas treatment. Because they operate at lower temperatures than PC boilers (760 to 930°C),

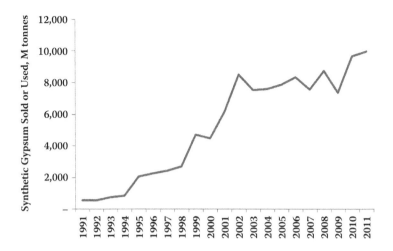

FIGURE 3.2 Synthetic gypsum sold or used in the United States (USGS Minerals Yearbook, 1991–2011).

NO_x formation is also minimized. An additional benefit of FBC technology is fuel flexibility, which allows for a variety of less expensive, lower quality fuels to be efficiently combusted.

3.3.1 POWER PRODUCTION WITH FBC

The earliest commercial versions of FBC technology were operated at atmospheric pressure, hence the term atmospheric fluidized bed combustion (AFBC). A later variation, pressurized fluidized bed combustion (PFBC), operates at elevated pressure and produces a high-pressure gas stream, which can be used to drive a gas turbine, while steam generated from water tubes in the chamber is used to drive a steam turbine, providing higher thermal efficiency with this combined cycle. Circulating fluidized bed combustion (CFBC) circulates large volumes of air and entrains sorbent material from the combustion chamber, where it is separated with large cyclones. Air is returned to fluidize the bed, while sorbent is also recycled to maximize sulfur capture. Spent sorbent is periodically removed and fresh sorbent is added as necessary to maintain emission requirements.

3.3.2 FBC MINERALOGY AND CHEMISTRY

The first step of sulfur capture in FBC is calcination of the sorbent, $CaCO_3$, followed by sulfation, as SO_2 evolved during combustion reacts with oxygen to form $CaSO_4$ (Koskinen et al.,1995):

$$CaCO_3 \text{ (s)} \rightarrow CaO \text{ (s)} + CO_2 \text{ (g)}$$

$$CaO \text{ (s)} + SO_2 \text{ (g)} + \tfrac{1}{2}O_2 \text{ (g)} \rightarrow CaSO_4 \text{ (s)}$$

At the higher pressures of PFBC, the high partial pressure of CO_2 prevents decomposition of $CaCO_3$ and sulfur capture occurs by direct sulfation:

$$CaCO_3(s) + SO_2(g) + \tfrac{1}{2}O_2(g) \rightarrow CaSO_4(s) + CO_2(g)$$

As with fly ash and bottom ash produced by pulverized coal combustion, the composition of ash generated by fluidized bed combustion can be quite variable. While fuel quality and composition determine a portion of the ash chemistry, sorbent composition is a major factor because it is such a large proportion of the by-product produced. Other factors such as operating conditions also contribute to ash variability as sorbent recycle and addition rates are adjusted to achieve the desired emission limits. A comparison of several FBC ash compositions is shown in Table 3.14. Note that when anthracite refuse was used as fuel, the by-product contained very high concentrations of SiO_2 and Al_2O_3 derived from the fuel, and lower concentrations of CaO and SO_3, presumably because of the lower amount of sorbent used. The other ashes are essentially similar in composition. It is particularly noteworthy that all of the ashes derived from high sulfur bituminous coal contain more than 5% SO_3, thus exceeding the specification limits for use as a pozzolan defined by ASTM C618 (ASTM, 2008a), and thus precluding use for this application.

As far as mineralogy is concerned, FBC ash is primarily comprised of anhydrite ($CaSO_4$), lime (CaO), quartz (SiO_2), and oxides of iron and magnesium as well as dehydroxylated clays derived from the ash component of the fuel. PFBC ashes also have been found to contain calcite ($CaCO_3$) and, in some cases, a distinct lack of lime (CaO) (Sellakumar et al., 1999).

TABLE 3.14

Comparison of Ash Chemistry from Different FBC Technologies

Technology	FBC	FBC	CFBC	PFBC	PFBC
	High Sulfur Bituminous		High Sulfur Bituminous	High Sulfur Bituminous	High Sulfur Bituminous
Fuel	Coal	Anthracite Refuse	Coal	Coal	Coal
SiO_2,%	26.05	56	24.4	21.84	29.46
Al_2O_3,%	10.59	21.24	9.56	9.68	12.48
Fe_2O_3,%	10.06	6.14	8.64	11.15	8.69
CaO,%	29.76	8.61	32.52	20.76	23.5
MgO,%	3.64	1.2	3.82	12.5	0.84
Na_2O,%	0.17	0.29	0.12	NA	1.07
K_2O,%	1.36	2.63	1.18	NA	1.27
P_2O_5,%	0.1	0.1	0.1	NA	0.5
TiO_2,%	0.45	1.35	0.41	NA	0.4
SO_3,%	15.42	3.6	18.21	10.58	20.83
LOI,%	9.13	NA	8.3	NA	0.26
Ref.	Robl et al., 2011	Robl et al., 2011	Stevens et al., 2009	Pflughoeft-Hassett, 1997	Bland et al., 1993

3.3.3 FBC Ash Utilization

3.3.3.1 Mining Applications

The major use of FBC ash in mining applications is as fill material for several specific purposes: subsidence prevention, acid mine drainage mitigation, reclamation soil neutralization, and cover soil stabilization. In the United States, many FBC units are located near abandoned mining and preparation facilities and utilize gob (a mixture of a small amount of coal with other mineral by-products from coal mining) as a fuel source. The abandoned sites frequently create environmental problems such as acidic mine drainage and unstable surface conditions; hence, FBC ash is extensively utilized for remediation. The free lime in FBC ash aids in neutralizing acidity, while self-cementing properties make it suitable for improving surface stability by improving soil cohesion.

Ash from FBC plants in Pennsylvania has been used in the reclamation of over 3400 acres of abandoned mine lands. When coal refuse piles are used as fuel, acid mine drainage and sediment pollution from these piles are eliminated and the alkaline ash is beneficially utilized for mine reclamation (Hornberger et al., 2005). The result has been the elimination of environmental and safety hazards, while improving water quality when waste material is recycled rather than landfilled. Similar beneficial results have been documented in coal mining areas of western Maryland, where different proportions of coal combustion by-products, including alkaline wet and dry FGD products, were injected as grout into abandoned mines. Post-injection water monitoring for over 8 years showed significant decreases in mine water acidity along with decreased iron and trace element concentrations in mine water (Bulusu et al., 2005).

3.3.3.2 Waste Stabilization

FBC ashes contain both free lime and anhydrite, making them effective for alkaline stabilization of sanitary sewage sludge. Hydration of lime to slaked lime and anhydrite to gypsum are both exothermic reactions. Heat generation, along with alkaline pH, provide pasteurization for elimination of pathogens and odor-causing bacteria. When mixed with sewage sludge, FBC-treated sludge produces a soil-like product with low odor that can readily be spread with conventional equipment for agriculture and reclamation.

3.3.3.3 Structural Fill/Road Base

FBC ash has been used in structural fills and road base application, much the same as fly ash and bottom ash. In China, FBC by-products have been used in highway construction applications such as road base stabilization, embankment construction, and as asphalt filler in surface courses (Li et al., 2010). The self-cementing properties of FBC ash offer an advantage in road base applications because smaller amounts of cementitious material are required to achieve final strength requirements. It is particularly important in these types of applications to ensure that proper prehydration occurs to minimize expansion. FBC materials formed in the boiler tend to have a predominance of anhydrite compared to gypsum. Since hydration of anhydrite to gypsum can cause expansion, it is important that sufficient water and time be allowed for prehydration to occur before final placement. Additionally, since FBC ashes also contain significant proportions of sulfur and lime, when combined with other cementitious compounds and water, ettringite can form. Ettringite is a cementitious mineral, but is highly expansive and potentially unstable. The combination of expansion and decomposition can cause poor structural performance. It is imperative that long-term durability be thoroughly characterized before FBC ash is used in any structural application. Many successful construction projects have been completed with FBC ash by prehydrating with sufficient water to convert anhydrite to gypsum and allow formation of primary ettringite to occur. In order for these expansive reactions to be completed, utilization must be delayed before final placement.

3.3.3.4   Agriculture

SDA by-products are used as a source of sulfur for agriculture in Germany, Denmark, and Austria (Jiang et al., 2011). FBC ash is often cited as a liming source to remedy acidic soils, along with providing a major source of Ca and S for plant nutrition (Terman et al., 1978; Stout and Priddy, 1996; Wang et al., 1994, 2006). Approximately half of the sulfur in slaked FBC ash is present as sparingly soluble ettringite, which can act as a slow release S fertilizer (Wang, 1996). PFBC ash, when compared to agricultural lime as amendment to acidic soils, has been found to support higher plant yield, possibly due to pH and nutritional effects (Brown et al., 1997). In addition, application of PFBC ash to sodic soils enhanced permeability.

FGD products enriched in calcium and sulfur can provide a source for these nutrients in agricultural applications. They can also control the transformation of nitrogen and phosphorus through immobilization and mineralization reactions (Seshadri et al., 2010).

3.3.3.5   Cement Manufacture

Despite self-cementitious properties, FBC ash cannot be used as a replacement for Portland cement in concrete because it does not meet ASTM C618 specifications. Excluding composition characteristics are typically deficiencies in SiO_2, Al_2O_3, and Fe_2O_3 with excessive SO_3. Depending on specific composition, it may be used as a raw material for clinker, and if sufficient gypsum is present, can be inter-ground with Portland clinker as a set retarder.

One area of cement manufacture where FBC ash has found application is in the manufacture of calcium sulfoaluminate (CSA) cement. Significant research and development has occurred in China on the manufacture of CSA cement from CFB by-products (Wang et al., 2005). CSA cement raw ingredients include CFB ash, fly ash, and limestone that are calcined at 1300 to 1350°C to produce calcium sulfoaluminate and dicalcium silicate (Wang et al., 2010). Advantages of CSA cements are rapid set, early high strength development, and low shrinkage, making them particularly well suited for pre-cast and repair applications. China's production of CSA cement in 1999 was estimated to be 1 million tonnes per year (Zhang and Glasser, 1999) and remains at that level currently (Edwards, 2011). CSA cements also have a long history of use in the United Kingdom for applications such as low shrinkage cement, shrinkage compensated slabs, mortar coatings for concrete pipes, rapid repair and setting mortars, anchor bolt grouts, and glass fiber reinforced cement products (Brown, 1993).

3.3.3.6   Biomass Ash

Biomass has become a fuel source of increasing interest worldwide, particularly when considering sustainable fuel resources. Biomass (e.g., energy crops, agricultural residue, trees, etc.) can be used as primary fuel, or co-fired with another fuel source such as coal. The by-product from these combustion processes (i.e., ash) does not conform to combustion ash specifications in the United States, and hence cannot be used as a pozzolan in concrete. Provisions have been made in Europe to allow biomass-derived ash to be used in cementitious applications, limited to when a maximum of 10% of the ash is derived from secondary fuel resources (European Standard EN-450, 2005).

The composition of ash derived from numerous biomass sources has been compiled by the Energy Research Centre of the Netherlands (ECN, 2011). While there is significant variation of composition between sources, biomass-derived ash is generally higher in alkali content than ash derived from coal and biomass is lower in ash content. The ash resulting from co-combustion of biomass and coal is comprised of proportional properties of the ash derived from both fuel components. Since biomass has low ash content and usually represents a minor proportion of the fuel mix, it typically has a minor impact on utilization properties. Co-combustion ash has been shown to be equivalent to ash produced from full coal firing with respect to environmental, technical, and occupational health properties, and even when derived from high proportions of biomass fuel, conforms to technical requirements of European standards (EN450) for utilization in concrete (te Winkel et al., 2007).

Ashes derived from utility co-combustion of coal with forest residues (i.e., tree limbs) and mill residue (i.e., sawdust) were evaluated for use in mortar and concrete by Shearer et al. (2011). While co-firing impacted morphology, no correlation was found between increased co-firing percentages and increased loss on ignition (LOI) or alkali content of the ash.

3.3.3.7   Utilization Statistics

Of the 9.31 million tonnes of FBC ash produced in the United States in 2010, 7.85 million tonnes were utilized in mining applications while a minor amount (65 kilo tonnes) was also used for waste stabilization (ACAA, 2010). In Europe (EU15), FBC ash production in 2008 was only approximately 1 million tonnes, with 0.2 million tonnes utilized, primarily in fill applications. Recent production and utilization data from other major coal utilizing countries is difficult to determine or even estimate. However, it is reasonable to assume that the amount of FBC by-products produced globally will increase significantly as this technology becomes more widely adopted, particularly for the combustion and co-combustion of biomass.

REFERENCES

Adholeya, A., Bhatia, N.P., Kanwar, S., and Kumar, S., 1998, Fly ash source and substrate for growth and sustainable agro-forestry system, Proceedings of Regional Workshop cum Symposium on Fly Ash Disposal and Utilization, organized by Kota Thermal Power Station, RSEB, Kota, Rajasthan, India.

Amaya, P.J., Booth, E.E., and Collins, R.J., 1997, Design and construction of roller compacted base courses containing stabilized coal combustion by-product materials.

Proceedings of the 12th International Symposium on Management and Use of Coal Combustion By-Products, Electric Power Research Institute, Report No. TR-107055, Vol. 1, Palo Alto, CA.

American Association of State Highway Officials (AASHO), 2011, Standard specifications for transportation materials and methods of sampling and testing, 31st Edition and AASHTO Provisional Standards, AASHTO Materials Reference Laboratory, Frederick, MD.

American Coal Ash Association (ACAA), 2010, 2010 Coal Combustion Product (CCP) Production & Use Survey Report, http://acaa.affiniscape.com/associations/8003/files/2010_CCP_Survey_FINAL_102011.pdf. Accessed January 3, 2010.

American Coal Ash Association (ACAA), 2003, Fly Ash Facts for Highway Engineers, U.S. Department of Transportation—Federal Highway Administration, Washington, D.C., National Technical Information Service, Springfield, Virginia.

American Coal Ash Association (ACAA), 1994, American Coal Ash Association Resource Bulletin RB-23, August, Aurora, CO.

American Society for Testing and Materials (ASTM), 2010, ASTM C331-00 Standard Specifications for Lightweight Aggregates for Concrete Masonry Units, Annual Book of ASTM Standards, Volume 04.02, ASTM International, West Conshohocken, PA.

American Society for Testing and Materials (ASTM), 2008a, ASTM C618, Standard specifications for coal fly ash and raw or calcined natural pozzolan for use as a mineral admixture in concrete, Annual Book of ASTM Standards, Volume 04.02, ASTM International, West Conshohocken, PA, pp. 330–332.

American Society for Testing and Materials (ASTM), 2008b, ASTM C595, Specification for blended hydraulic cements, Annual Book of ASTM Standards, Volume 04.01, ASTM International, West Conshohocken, PA.

American Society for Testing and Materials (ASTM), 2008c, ASTM D5239-04 Standard practice for characterizing fly ash for use in soil stabilization, Annual Book of ASTM Standards, Volume 04.08, ASTM International, West Conshohocken, PA.

American Society for Testing and Materials (ASTM), 2008d, ASTM D698-07e1 Standard test methods for laboratory compaction characteristics of soil using standard effort (12 400 ft-lbf/ft^3 (600 kN-m/m^3)), Annual Book of ASTM Standards, Volume 04.08, ASTM International, West Conshohocken, PA.

Appleyard, F.C., 1975, Gypsum and anhydrite, in *Industrial Minerals,* 4th ed, S.J. Lefonde (Ed.), American Institute of Mining, Metallurgical and Petroleum Engineers, pp. 185–199.

Araujo, J.H., da Silva, N.F., Acchar, W., and Gomes, U.U., 2004, Thermal decomposition of illite, *Mater. Res.,* 7(2).

Ash Development Association of Australia (ADAA), 2010, Annual Membership Survey Results, January–December 2010, HBM Pty Ltd., http://www.adaa.asn.au/documents/ADAA_Mship_Report_2010.pdf_(accessed March 20, 2012).

Association of Canadian Industries Recycling Coal Ash (CIRCA), 2006, Production and Use of Coal Combustion Products, http://www.circainfo.ca/documents/ProductionandUseStatistics.pdf (accessed January 20, 2012).

Beeghly, J.H. and Schrock, M., 2009, Dredge material stabilization using the pozzoalnic or sulfo-pozzolanic reaction of lime by-products to make an engineered structural fill, *Proceedings World of Coal Ash Conference*, Lexington, KY, May 4–7.

Bengtsson, S. Semi-dry FGD end product utilization—European experiences, presented at the Mega Symposium, Chicago, IL, August 20–23, 2001; Paper 219.

vom Berg, W., Peters, F., Puch, K.-H., and Taubert, U., 1993, Use of residues from coal-fired power stations in the Federal Republic of Germany, UNIPEDE/IEA Conference on Thermal Power Generation and the Environment (Paper 933en3.6), Hamburg, Germany, September 1–3.

Bland, A.E., Brown, T.H., and Sellakumar, K.M., 1993, The use of PFBC ash to remediate acidic and sodic soil conditions, Final Report, DE-FC21-93MC30127, U.S. Department of Energy, Federal Energy Technology Center, Pittsburgh, PA, http://uwlib5.uwyo.edu/omeka/archive/files/the-use-of-pfbc-ash-to-remediate-acidic-and-sodic-soil-conditions_abec2388e4.pdf (accessed April 12, 2012).

Bloss, W., 1984, Process with Calcium Sulfite Hemihydrate in a Powdery Byproduct from Dry Flue Gas Desulfurization for the Production of Fly Ash Cement. U.S. Patent 4,470,850.

Brendel, G.F. and Glogowski, P.E., 1989, Ash utilization in highways: Pennsylvania demonstration project, Electric Power Research Institute, Report No. GS-6431, Palo Alto, CA.

Brendel, G.F., Balsamo, N.J., and Wei, L.H., 1997, Guidelines for the Beneficial Use of Advanced SO2 Control By-Products, TR-108403, Electric Power Research Institute, Palo Alto, CA.

Brown, T.H., Bland, A.E., and Wheeldon, J.M., 1997, Pressurized fluidized bed combustion ash 2. Soil and mine soil amendment use options, *Fuel*, 76(8): 741–748.

Brown, A.D.R., 1993, Concrete 2000: Economic and Durable Construction Through Experience, Proceedings of the International Conference, Dihr, R.K. and Dyer, T.D., Eds., University of Dundee, Dundee, Scotland.

Bulusu, S., Aydilek, A.H., Petzrick, P., and Guynn, R., 2005, Remediation of abandoned mines using coal combustion by-products, *Journal of Geotechnical and Geoenvironmental Engineering*, 131(8): Aug. 1.

Burns, H. and Gremminger, L., 1994, Lime and fly ash stabilization of wastewater treatment sludge, U.S. Patent 5,277,826.

Bye, G.C., 1999, *Portland Cement*, 2nd ed., London: Thomas Telford, 206–208.

Cao, D.-Z., Selic, E., and Herbell, J.-D., 2008, Utilization of fly ash from coal-fired power plants in China, *Journal of Zhejiang University SCIENCE A*, 9(5): 681–687.

Capp, J.P., 1978, Power plant fly ash utilization for land reclamation in the eastern United States, in: *Reclamation of Drastically Disturbed Lands*, Schaller, F.W. and Sutton, P., Eds., ASA, Madison, WI, 339–353.

Carlson, C.L. and Adriano, D.C., 1993, Environmental impacts of coal combustion residues, *J. Environ. Qual.*, 22: 227–247.

Ceratech, 2012, http://www.ceratechinc.com/ (accessed March 14, 2012).

Chang, A.C., Lund, L.J., Page, A.L., and Warneke, J.E., 1977, Physical properties of amended soils, *J. Environ. Qual.*, 6(3): 267–270.

Chen, L. and Dick, W., 2011, Gypsum as an agricultural amendment: general use guidelines, The Ohio State University Extension, Bulletin 945, Columbus, OH.

Chen, L., Kost, D., and Dick, W., 2009, Reclamation of abandoned surface coal mine lands using flue gas desulfurization products, *Energeia*, 20(5), University of Kentucky Center for Applied Energy Research, Lexington, KY.

Coal Mining in Ukraine, 2009, General profile-market study, http://www.coalukraine.com/coalminingukraine/generalprofile.html (accessed January 20, 2012).

Connell, D., 2009, Greenridge multi-pollutant control project, DOE Final Report, Cooperative Agreement No. DE-FC26-06NT41426, April.

Dawson, G., Perri, J.S., and Daley, J.R., 1987, Utilization potential of advanced SO$_2$ control by-products, EPRI-CS-5269, Electric Power Research Institute, Palo Alto, CA.

Davidovits, J. and Comri, D., 1988, Long term durability of hazardous toxic and nuclear waste disposal, Proceedings of Geopolymer '88, Compiegne, France.

Davidovits, J., 1994, Properties of geoploymers cements, Proceedings 1st International Conference on Alkaline Cements and Concretes, pp. 131–149.

Deer, W.A., Howie, R.A., and Zussman, J., 1992, *An Introduction to the Rock Forming Minerals*, 2nd ed., Pearson Education Ltd., England.

Dick, W., Bigham, R., Forster, F. et al., 1999, Land application uses for dry flue gas desulfurization by-products, United States Geological Survey, Ohio State University, Wooster, OH.

Duo, W., Dam-Johansen, K., and Østergaard, K., 1992, Kinetics of the gas-phase reaction between nitric oxide, ammonia and oxygen, *Can. J. Chem. Engineering*, 70: 1014–1020.

Edwards, P., 2011, Future cement-looming beyond OPC, *Global Cement Magazine*, http://www.globalcement.com/magazine/articles/315-future-cement-looking-beyond-opc (accessed April 13, 2012).

Energy Information Agency (EIA), 2011, Annual Energy Outlook 2011, Report #:DOE/EIA-0383(2011), National Energy Information Center, Energy Information Administration, Washington, DC.

Energy Research Center of the Netherlands (ECN), 2011, Phyllis, database for biomass and waste, http://www.ecn.nl/phylli_(accessed May 1, 2012).

Eskom, 2010, Ash Management in Eskom Factsheet, http://eskomtest0.ensightnetworkscluster0.com/content/CO_0004AshManRev7~1.pdf (accessed February 1, 2012).

European Coal Combustion Products Association (ECOBA), 2011, Joint EURELECTRIC/ECOBA briefing: classification of coal combustion products under revised waste framework directive (2008/98/EC), http://www.eurelectric.org/Download/Download.aspx?DocumentFileID=69075 (accessed May 20, 2012).

European Standard EN-450, 2005, Fly ash for concrete-Part 1: definition, specifications and conformity criteria, Brussels, Belgium.

Fail, J.L. and Wochok, Z.S., 1977, Soyabean growth on fly ash amended strip mine soils, *Plant and Soil*, 48: 473–484.

Farber, P.S., Livengood, C.D., and Anderson, J.L., 1983, Leachate of dry scrubber wastes, presented at the 76th Annual Meeting of the Air Pollution Control Agency, Atlanta, GA, June 19–24, 1983.

Farina, M.P.W. and Channon, P., 1988, Acid-subsoil amelioration: II. Gypsum effects on growth and subsoil chemical properties, *Soil Science Society of America Journal*, 52: 175–180.

Fuerborn, H.J., 2011, Coal combustion products in Europe—an update on production utilization, standardization and regulation, Proceedings World of Coal Ash, Denver, CO, May 9–12.

Feuerborn, J., 2010, CCPs in Europe—production, quality and use today and tomorrow, presentation at EuroCoalAsh 2010, May 27–28, Copenhagen Denmark, http://www.ecoba.com/evjm,media/EUROCOALASH/01_Feuerborn.pdf (accessed March 20, 2012).

Gaikwad, R., 2002, Dry flue gas desulfurization technology evaluation, Sargent and Lundy Project 11311-000, Chicago, IL, prepared for National Lime Association, http://www.graymont.com/technical/Dry_Flue_Gas_Desulfurization_Technology_Evaluation.pdf (accessed April 2, 2012).

Gluskoter, H.J., Ruch, R.R., Miller, W.G. et al., 1977, Trace elements in coal: occurrence and distribution, Illinois State Geological Survey Circular 499, Urbana, IL.

Guynn, R., Litten, J. et al., 2009, Use of a CCB grout barrier to reduce the formation of acid mine drainage: the siege of acre project, Kempton, Maryland, Proceedings 2009 World of Coal Ash Conference, Lexington, KY, May 4–7.

Haering, K.C. and Daniels, W.L., 1991, Fly ash: characteristics and use in mineland reclamation. A literature review, *Virg. Coal Energy J.*, 3: 33–46.

Haseltine, B.A., 1975, Comparison of energy requirements or building materials and structures, *The Structural Engineer*, 53(9): 357–365.

Heebink, L.V., Buckley, T.D., Hassett, D.J. et al., 2007a, Current status of spray dryer absorber material characterization and mineralogy, Proceedings 2007 World of Coal Ash, Covington, KY, May 7–10.

Heebink, L.V., Buckley, T., Pflughoeft-Hassett, D., and Hassett, D., 2007b, A review of literature related to the use of spray dryer absorber material: production, characterization, utilization applications, barriers, and recommendations, EPRI, Palo Alto, CA and UND EERC CARRC, Grand Forks, ND, Technical Report 1014915.

Heyliger, P.R., Allen, D., Lebsack, M., and Wilson, T., 2011, Feasibility study for highway traffic noise barriers from a spray dryer ash and used rubber composite, Final Report, Civil and Environmental Engineering Department, Colorado State University, Ft. Collins, CO, http://www.cdphe.state.co.us/oeis/p2_program/grantreports/atgcsuhwy.pdf (accessed March 14, 2012).

Hornberger, R.J., Dalberto, A.D., Menghini, M.J. et al., 2005, Coal ash beneficial use at mine sites in Pennsylvania, Proceedings 2005 World of Coal Ash (WOCA), Lexington, KY, April 11–15.

Hower, J.C. and Parekh, B.K., 1991, Chemical/physical properties and marketing, in: *Coal Preparation,* 5th ed., Leonard, J., Ed., Society for Mining, Metallurgy and Exploration, Littleton CO, 1–94.

Hower, J.C., Rathbone, R.F., Graham, U.M., Groppo, J.G., Brooks, S.M., Robl, T.L., and Medina, S.S., 1995, Approaches to the petrographic characterization of fly ash, Proceedings of 11th International Coal Testing Conference, Lexington, KY, May 10–12.

Hower, J.C., Rathbone, R.F., Robertson J.D., Peterson, G., and Trimble, A.S., 1999a, Petrology, mineralogy, and chemistry of magnetically-separated sized fly ash, *Fuel,* 78(2):197–203.

Hower, J.C., Maroto-Valer, M.M., Taulbee, D.N., and Sakulpitakphon, T., 1999b, Mercury capture by distinct fly ash carbon forms, *Energy Fuels,* 14(1): 224–226.

International Energy Agency (IEA), 2011, Power generation from coal, ongoing developments and outlook, Information Paper, http://www.iea.org/papers/2011/power_generation_from_coal.pdf (accessed January 12, 2012).

JCOAL, Japan Coal Energy Center, 2006, Utilization of coal ash, http://www.jcoal.or.jp/coaltech_en/coalash/ash02e.html (accessed February 2012).

Jaarsveld, J.G., Deventer, J.S., and Lorenzen, L., 1998, Factors affecting the immobilization of metals in geopolymerised fly ash, *Metallurgical and Materials Transactions B,* 29: 659–669.

Jala, S. and Goyal, D., 2006, Fly ash as a soil ameliorant for improving crop production-a review, *Biosource Technology,* 97(9): 1136–1147.

Jastrow, J.D., Zimmerman, A.J., Dvorak, A.J., and Hinchman, R.R., 1981, Plant growth and trace element uptake on acidic coal refuse amended with lime or fly ash, *J. Environ. Qual.,* 10: 154–160.

Jiang, Y., Wu, M-C.W, Su, Q. et al., 2011, Dry CFBC-FGD by-product utilization-international perspectives, Proceedings World of Coal Ash (WOCA) Conference, Denver, CO, May 9–12.

Kim, J., Cho, S., Kong, J. et al., 2011, Hardening characteristics of controlled low strength material made of coal ash, Proceedings World of Coal Ash, Denver, CO, May 9–12.

Kolar, J., 1995, Possibilities of using residual products of the spray absorption processes, *VGB Kraftwerkstechnik,* 2: 153–159.

Koskinen, J., Lehtonen, P., and Sellakumar, K.M., 1995, Ultraclean combustion of coal in pyroflow PCFB combustors, Proceedings of the 13th International Conference on Fluidized Bed Combustion, Orlando, FL, pp. 369–378.

Koslowski, T. and Roggendorf, H., 1996, Binder for interior plasters, U.S. Patent 5,522,928.

Krishna, M.G., 1980, Indian coal industry: past, present and future, in: *Energy Policy for India,* Pachauri, R.K., Ed., Macmillian Company of India, Delhi.

Lamminen, M., Wood, J., Walker, H., and Chin, Y.P., 1999, Acid mine drainage abatement using flue gas desulfurization byproduct: water quality aspects, Proceedings 1999 International Ash Utilization Symposium, Lexington, KY.

Lee, J.C., Bradshaw, S.L., Edil, T.B., and Benson, C.H., 2011, Quantifying the benefits of flue gas desulfurization gypsum in sustainable wallboard production, Proceedings World of Coal Ash Conference, Denver, CO, May 9–12.

Lea, F.M., 1970, *The Chemistry of Cement and Concrete,* Edward Arnold Ltd., London.

Lee, R., Rafalko, L., and Giacinto, J., 2008, 10-year update on the Winding Ridge Project, prepared for Maryland Power Plant Research Program, PPRP-147.

LifeTime Composites, 2012, http://bluelinxco.com/Portals/0/8%20Page%20Distributor%20 Brochure.pdf (accessed March 13, 2012).

Lokhande, R.D., Prakash, A., Singh, K.B., and Singh, K.K., 2005, Subsidence control measures in coalmines: A review, *Journal of Scientific and Industrial Research*, 64: 323–332.

Li, C., Su, D., and Zhang, J., 2010, Research on preparation of bitumen filler by desulfurization residue and limestone powder, *Journal of Qindao Technological University*, 31(1): 204–210.

Luna, Y., Fernandez-Pereira, C., Vale, J.F., and Alberca, L., 2009, Waste stabilization/solidification (S/S/) using fly ash-based geopolymers. Influence of carbonation on the S/S of an EAF dust, Proceedings 2009 World of Coal Ash, May 4–7, Lexington, KY.

Lyons, C. and Petzrich, P., 1995, The Maryland haulback project-the use of coal combustion by-products to abate acid mine drainage in a Maryland underground coal mine: a demonstration project, Proceedings 1999 International Ash Utiliz. Symposium, Lexington, KY.

Malhotra, V.M., 1986, Role of supplementary cementing materials in reducing greenhouse gas emissions, Materials Technology Laboratory, CANMET, Internal Report.

Manz, O.E., 1997, Worldwide production of coal ash and utilization in concrete and other products, *Fuel*, 76(8): 691–696.

Mastalerz, M., Hower, J.C., Drobniak, A., Mardon, S.M., and Lis, G., 2004, From in-situ coal to fly ash: a study of coal mines and power plants from Indiana, *International Journal of Coal Geology*, 59 (1–3): 171–192.

Meiers, R.J., Golden, D., Gray, R., and Yu, W.C., 1995, Fluid placement of fixated scrubber sludge to reduce surface subsidence and to abate acid mine drainage in abandoned underground coal mines, Proceedings 1999 International Ash Utiliz. Symposium, Lexington, KY.

Meij, R., 1994, Trace element behavior in coal-fired power plants, *Fuel Processing Technology*, 39(1–3): 199–217.

Murakami, K., 1968, Utilization of chemical gypsum for Portland cement, ISCC Session 5, Part IV, p. 466, Tokyo.

Namagga, C.A., 2010, Valuable utilization of spray dryer ash and its performance in structural concrete, M.S. Thesis, Department of Civil and Environmental Engineering, Colorado State University, Ft. Collins, CO.

Nelson, J.B., 1953, Assessment of the mineral species associated with coal, *Monthly Bulletin, British Coal Utilization Research Association*, 17(2): 41–45.

Norton, L.D., Shainberg, I., and King, K.W., 1993, Utilization of gypsiferous amendments to reduce surface sealing in some humid soils of the eastern USA, in: *Soil Surface Sealing and Crusting, Catena Supplement*, Poesen, J.W.A. and Newaring, M.A., Eds., 24: 77–92.

Nugteren, H.W., 2007, Coal fly ash: from waste to industrial product, *Part. Part. Syst. Charact.*, 24(1): 49–55.

Parsa, J., Munson-McGee, S.H., and Steiner, R., 1996, Stabilization/solidification of hazardous wastes using fly ash, *J. Environ. Eng.*, 122(10).

Pasini, R.A., 2009, An evaluation of flue gas desulfurization gypsum for abandoned mine land reclamation, M.S. Thesis, The Ohio State University, Columbus, OH.

Pflughoeft-Hassett, D., 1997, IGCC and PFBC by-products: generation, characteristics, and management practices, Final Report for Environmental Protection Agency, Office of Solid Waste, 97-EERC-09-06.

Phung, H.T., Lam, H.V., Lund, L.J., and Page, A.L., 1979, The practice of leaching boron and slats from fly ash amended soils, *Water, Air Soil Pollution*, 12: 247–254.

Portland Cement Association, 2012, History & manufacture of Portland cement, http://www. cement.org/basics/concretebasics_history.asp (accessed January 23, 2012).

Prusinski, J.R., Cleveland, M.W., and Saylak, D., 1995, Development and construction of road bases from flue gas desulfurization material blends, Proceedings 11th International Ash Utilization Symposium, Electric Power Research Institute, Report No. TR-104657, Vol. 1, Palo Alto, CA.

Putilov, Y.V. and Putilova, I., 2005, Modern approach to the problem of utilization of fly ash and bottom ash from power plants in Russia, Proceedings 2005 World of Coal Ash, Lexington, KY, April 11–15.

Rajendran, N., 1994, Controlled low strength materials (CLSM), ACI Committee 229 Report, http://www.osti.gov/bridge/servlets/purl/505263-DMiGgX/webviewable/505263.pdf (accessed January 23, 2012).

Reddi, L.N., Rieck, G.P., Schwab, A.P., Chou, S.T., and Fan, L.T., 1996, Stabilization of phenolics in foundry waste using cementitious materials, Journal of Hazardous Materials, 45(2): 89–106(18).

Reeves, G., 2009, The use of coal combustion products by local watershed groups to seal stream loss zones in karst topography, Proceedings 2009 World of Coal Ash Conference, Lexington, KY, May 4–7.

Riekerk, H., 1984, Coal-ash effects on fuelwood production and runoff quality, Southern J. Appl. For., 8: 99–102.

Robl, T., Oberlink, A., Brien, J., and Pagnotti, J., 2011, The utilization potential of anthracite CFBC spent bed fly ash as a concrete additive, Proceedings 2011 World of Coal Ash (WOCA) Conference, Denver, CO, May 9–12.

Scott, W.D., McCraw, B.D., Motes, J.E., and Smith, M.W., 1993, Application of calcium to soil and cultivar affect elemental concentration of watermelon leaf and rind tissue, Journal of the American Society of Horticultural Science, 118: 201–206.

Sellakumar, K.M., Conn, R., and Bland, A., 1999, A comparison study of ACFB and PCFB ash characteristics, Proceedings of 6th International Conference on Circulating Fluidized Beds, Wurzburg, Germany, Aug. 22–27.

Seshadri, B., Bolan, N.S., Naidu, R., and Brodie, K., 2010, The role of coal combustion products is managing the bioavailability of nutrients and heavy metals in soils, Journal of Soil Science and Plant Nutrition, 10(3): 378–398.

Sezer, A., Inan, G., Yilmaz, H.R., and Ramyar, K., 2006, Utilization of a very high lime fly ash for improvement of Izmir clay, Building and Environment, 41(2): 150–155.

Shainberg, I., Sumner, M. E., Miller W.P., et al., 1989, Use of gypsum on soils. A review, Advances in Soil Science, 9: 1–111.

Shear, C.B., Ed., 1979, International Symposium on Calcium Nutrition of Economic Crops Communications, New York: Marcel Dekker Publishing.

Shearer, C.R., Yeboah, N., Kustis, K.E., and Burns, S.E., 2011, The early age behavior of biomass fired and co-fired fly ash in concrete, Proceedings 2011 World of Coal Ash (WOCA) Conference, Denver, CO, May 9–12.

Sing, R.N., Keefer, R.F., Gazhi, H.E., and Horvath, D.J., 1982, Evaluation of plant growth and chemical properties of mine soils following application of fly ash and organic waste materials, Proceedings 6th Symposium America Coal Ash Association, Reno, NV.

Singh, Y., 2010, Fly ash utilization in India, http://www.wealthywaste.com/fly-ash-utilization-in-india_(accessed February 1, 2012).

Smith, C.L., 1985, FGD sludge C coal ash road base: seven years of performance, Proceedings of the 8th International Coal and Solid Fuels Utilization Conference, Pittsburgh, PA, November.

Smith, C.L., 1989, The first 100,000 tons of stabilized scrubber sludge in roadbase construction, Proceedings of the Power-Gen '89 Conference, New Orleans, LA, December.

Smyth, T.J. and Cravo, S., 1992, Aluminum and calcium constraints to continuous crop production in a Brazilian Amazon soil, Agronomy Journal, 84: 843–850.

Spackman, W. and Moses, R.G., 1961, The nature and occurrence of ash-forming minerals in anthracite, Bulletin 75, Mineral Industries Experiment Station, Pennsylvania State University, pp. 1–15.

Stevens, W., Robl, T., and Mahboub, K., 2009, The cementitious and pozzolanic properties of fluidized bed combustion fly ash, Proceedings 2009 World of Coal Ash (WOCA) Conference, Lexington, KY, May 4–11.

Stout, W.L. and Priddy, W.E., 1996, Use of flue gas desulphurisation (FGD) by-product gypsum on alfalfa, *Commun. Soil Sci. Plant Anal.*, 27: 2419.

Swan, C., Topping, G., and Kashi, M.G., 2007, Flowable fills developed with high volumes of fly ash, Proceedings 2007 World of Coal Ash, Northern Kentucky, May 7–10.

Taylor, H.F.W., 1990, *Cement Chemistry*, London: Academic Press, 10–11.

Teichmöller, M. and Teichmöller, R., 1982, The geological basis of coal formation, in: *Coal Petrology*, 3rd ed., Stach, E. et al., Eds., pp. 5–86.

Terman, G.L., Kilmer, V.J., Hunt, C.M., and Buchanan, W., 1978, Fluidized bed boiler waste as a source of nutrients and lime, *J. Environ. Qual.*, 7: 147–151.

te Winkel, H., Meij, R., and Saraber, A., 2007, Environmental and health aspects of ashes produced at co-combustion of biomass, Proceedings 2007 World of Coal Ash (WOCA), Northern Kentucky, May 7–10.

Toma, M., Sumner, M.E., Weeks, G., and Saigusa, M., 1999, Long-term effects of gypsum on crop yield and subsoil chemical properties, *Soil Science Society of America Journal*, 63: 891–895.

United States Environmental Protection Agency (USEPA), 2008, Using Recycled Industrial Materials in Buildings, U.S. Environmental Protection Agency EPA530-F-08-022, October 2008, http://www.epa.gov/osw/conserve/rrr/imr/pdfs/recy-bldg.pdf (accessed March 14, 2012).

United States Federal Highway Administration (USFHWA), 1997, User Guidelines for Waste and Byproduct Materials in Pavement Construction, Publication Number: FHWA-RD-97-148, http://www.fhwa.dot.gov/publications/research/infrastructure/structures/97148/fgd1.cfm.

United States Geological Survey (USGS) Minerals Yearbooks, 1991–2011, Gypsum, U.S. Department of Interior, http://minerals.usgs.gov/minerals/pubs/commodity/gypsum/ (accessed March 20, 2012).

Uomoto, T. and Kobayashi, K., 1983, 1st International Conference on Fly Ash, Silica Fume, Slag and Other Mineral By-Products in Concrete, ACI, SP-79, Detroit, Vol. 2, p. 1013.

van de Lindt, J.W. and Rechan, R.K., 2011, Seismic performance comparison of a high-content SDA frame and standard RC frame, *Advances in Civil Engineering*, Article ID 478475.

Visuvasam, D., Selvaraj, P., and Sekar, S., 2005, Influence of coal properties on particulate emission control in thermal power plants in India, Second International Conference on Clean Coal Technologies for Our Future (CCT 2005), Sardinia, Italy.

Wang, H.L., 1996, Potential uses of fluidized bed boiler ash (FBA) as a liming material, soil conditioner and sulphur fertilizer, PhD thesis, Massey University, Palmerston North, New Zealand.

Wang, H., Bolan, N.S., Hedley, M.J., and Home, D.J., 2006, Potential uses of fluidized bed boiler ash (FBA) as a liming material, soil conditioner and sulfur fertilizer, in: *Coal Combustion By-Products and Environmental Issues*, Sajwan, K.S., Twardowska, I., Punshon, T., and Alva, A.K., Eds., Springer Publishers, New York, 202–215.

Wang, W.-L., Cui, L., and Ma, C-Y., 2005, Study on properties and comprehensive utilization of dry and semi-dry desulfurization residues, *Power Systems Engineering*, 21(5): 27–29.

Wang, H.L., Hedley, M.J., and Bolan, N.S., 1994, Chemical properties of fluidised bed boiler ash relevant to its use as a liming material and fertilizer, *NZ J. Ag. Res.*, 38: 249–256.

Wang, W.-L., Wang, P., Ma, C., and Luo, H., 2010, Calculation for mineral phases in the calcination of desulfurization residue to produce sulfoaluminate cement, *Industrial and Engineering Chemistry Research*, 40: 9504–9510.

Ward, C., French, D., Jankowski, J., Riley, K., and Li, Z., 2006, Use of coal ash in mine backfill and related applications, Research Report 62, Cooperative Centre for Coal in Sustainable Development, CQAT Technology Transfer Centre, Pullenvale, Queensland, Australia.

Ward, C.R. and French, D., 2005, Relationship between coal and fly ash mineralogy, based on quantitative x-ray diffraction methods, Proceedings 2005 World of Coal Ash, Lexington, KY, April 11–15.

Webster, W.C., 1982, Utilization of dry scrubber waste materials. U.S. Patent 4,354,876, October 19.

Weisenstein, L.H., Osmeloski, J.F., Rutzke, M. et al., 1989, Elemental analysis of grasses and legumes growing on soil covering coal ash landfill sites, *Journal of Food Safety*, 9: 291–300.

Wolfe, W., Butalia, T., Walker, H., and Baker, R., 2009, Befeficial use of FGD by-products in mine land reclamation, Proceedings 2009 World of Coal Ash, Lexington, KY, May 4–7.

World Coal Association, 2011, Coal statistics, http://www.worldcoal.org/resources/coal-statistics/ (accessed December 20, 2011).

Wu, Z. and Naik, T., 2003, Chemically activated blended cements, *Materials Journal*, 100: 434–440.

Zhang, L. and Glasser, F.P., 1999, Modern concrete materials: binders, additions and admixtures, Proceedings of the International Conference, R.K. Dihr and T.D. Dyer, Eds., University of Dundee, Dundee, Scotland, pp. 261–274.

Zhang, J., Provis, J.L., Feng, D., and van Deventer, J., 2008, Geopolymers for immobilization of Cr^{6+}, Cd^{2+}, and Pb^{2+}, *Journal of Hazardous Materials*, 157(2–3): 587–598.

Zhang, L., Su, M., and Wang, Y., 1999, Development of the use of sulfo- and ferroaluminate cements in China, *Advances in Cement Research*, 11(1): 15–21.

4 Waste Electrical and Electronic Equipment (WEEE)

I.D. Pulford

CONTENTS

Waste electrical and electronic equipment (WEEE) is a rapidly growing category of waste material, which has been of particular concern over the last decade due to risks to human health and environmental contamination resulting from some of the toxic components used in its manufacture. This has been especially the case in countries of the Organisation for Economic Cooperation and Development (OECD), where the market for such goods is continually inundated with new models and therefore, turnover rates, and hence disposal rates, are high. Legislation has been quickly put in place in many of these countries, setting targets for collection, recycling, and disposal of WEEE. Developing countries are rapidly catching up in terms of WEEE use, and some have put legislation in place. However, these countries are faced with the problems of the importing of old equipment, which often contains higher amounts of toxic components, and the informal dismantling of equipment with no health or environmental controls.

The opportunity for reuse of WEEE lies not in conversion into a useful product, but in the recovery of economically valuable components. Much attention is paid to recovery of high-value metals, such as gold, copper, and palladium, but in order to meet legislative targets for recycling WEEE, other components such as plastics and ferrous metals are also being recovered.

4.1   LEGISLATION DEALING WITH WEEE

The introduction of legislation dealing specifically with WEEE has been increasing over the last 15 years as the scale of the problem becomes more apparent. Switzerland was the first country to have legislation specifically aimed at WEEE (1998), but in many countries, there is still reliance on non-WEEE–specific measures, such as the Basel Convention on the Control of Transboundary Movements of Hazardous Wastes and their Disposal. Summaries of worldwide WEEE legislation have been recently published (Ongondo et al., 2011; de Oliveira et al., 2012), but this is, of course, a continually changing situation. In recent years, many developed countries have put in place some type of regulation on the disposal of WEEE as the amount of products sold, and hence waste, has increased.

4.1.1   EUROPEAN UNION (EU)

The countries of the European Union (EU) have been very much at the forefront in setting up a legislative framework for dealing with WEEE. Since 2003, two directives have covered the production and disposal of WEEE products within the EU.

- *The Restriction of Use of Certain Hazardous Substances (RoHS) Directive (2002/95/EC)* limited the amounts of the heavy metals cadmium, hexavalent chromium, lead, and mercury, and the polybrominated biphenyls (PBBs) and polybrominated diphenyl ether (PBDE) flame-retardants. The aim was to reduce the health risks and the impact on the environment when products containing these toxic substances are eventually dismantled and disposed of.
- *Directive 2002/96/EC* defined classes of WEEE (Table 4.1) and provided the legislative framework for the collection, recycling, and disposal of WEEE, and has recently (May 2012) been amended (http://register.consilium. europa.eu/pdf/en/12/pe00/pe00002.en12.pdf). This amendment is aimed at all those involved with the production, distribution, and use of electrical and electronic equipment. From 2016, EU member states will be required to collect 45% (on a weight basis) of the electrical and electronic equipment sold in their national markets, and this figure will rise to 65% from 2019. There are provisions for free collection points in retail outlets for small items to encourage collection of old equipment. There is also legislation to limit the export of WEEE from the EU to developing countries. The focus of this directive is to promote reuse and recycling.

TABLE 4.1

Categories of WEEE Defined in EU Directive 2002/96/EC, Amended May 2012

1. Large household appliances	Large cooling appliances
	Refrigerators
	Freezers
	Other large appliances used for refrigeration, conservation, and storage of food
	Washing machines
	Clothes dryers
	Dish washing machines
	Cookers
	Electric stoves
	Electric hot plates
	Microwaves
	Other large appliances used for cooking and other processing of food
	Electric heating appliances
	Electric radiators
	Other large appliances for heating rooms, beds, seating furniture
	Electric fans
	Air conditioner appliances
	Other fanning, exhaust ventilation, and conditioning equipment
2. Small household appliances	Vacuum cleaners
	Carpet sweepers
	Other appliances for cleaning
	Appliances used for sewing, knitting, weaving, and other processing for textiles
	Irons and other appliances for ironing, mangling, and other care of clothing
	Toasters
	Fryers
	Grinders, coffee machines, and equipment for opening or sealing containers or packages
	Electric knives
	Appliances for hair-cutting, hair drying, tooth brushing, shaving, massage, and other body care appliances
	Clocks, watches, and equipment for the purpose of measuring, indicating, or registering time
	Scales
3. IT and telecommunications equipment	Centralized data processing:
	Mainframes
	Minicomputers
	Printer units

(Continued)

TABLE 4.1 (*Continued*)
Categories of WEEE Defined in EU Directive 2002/96/EC,
Amended May 2012

	Personal computers (CPU, mouse, screen, and keyboard included)
	Laptop computers (CPU, mouse, screen, and keyboard included)
	Notebook computers
	Notepad computers
	Printers
	Copying equipment
	Electrical and electronic typewriters
	Pocket and desk calculators and other products and equipment for the collection, storage, processing, presentation, or communication of information by electronic means
	User terminals and systems
	Facsimile machine (fax)
	Telex
	Telephones
	Pay telephones
	Cordless telephones
	Cellular telephones
	Answering systems and other products or equipment of transmitting sound, images, or other information by telecommunications
4. Consumer equipment and photovoltaic panels	Radio sets
	Television sets
	Video cameras
	Video recorders
	Hi-fi recorders
	Audio amplifiers
	Musical instruments and other products or equipment for the purpose of recording or reproducing sound or images, including signals or other technologies for the distribution of sound and image other than by telecommunications
	Photovoltaic panels
5. Lighting equipment	Luminaries for fluorescent lamps with the exception of luminaries in households
	Straight fluorescent lamps
	Compact fluorescent lamps
	High-intensity discharge lamps, including pressure sodium lamps and metal halide lamps
	Low-pressure sodium lamps
	Other lighting or equipment for the purpose of spreading or controlling light with the exception of filament bulbs

(*Continued*)

TABLE 4.1 (*Continued*)
Categories of WEEE Defined in EU Directive 2002/96/EC, Amended May 2012

6. Electrical and electronic tools (with the exception of large-scale stationary industrial tools)	Drills
	Saws
	Sewing machines
	Equipment for turning, milling, sanding, grinding, sawing, cutting, shearing, drilling, making holes, punching, folding, bending, or similar processing of wood, metal, and other materials
	Tools for riveting, nailing, or screwing or removing rivets, nails, screws, or similar uses
	Tools for welding, soldering, or similar use
	Equipment for spraying, spreading, dispersing, or other treatment of liquid or gaseous substances by other means
	Tools for mowing or other gardening activities
7. Toys, leisure, and sports equipment	Electric trains or car racing sets
	Hand-held video game consoles
	Video games
	Computers for biking, diving, running, rowing, etc.
	Sports equipment with electric or electronic components
	Coin slot machines
8. Medical devices (with the exception of all implanted and infected products)	Radiotherapy equipment
	Cardiology equipment
	Dialysis equipment
	Pulmonary ventilators
	Nuclear medicine equipment
	Laboratory equipment for in-vitro diagnosis
	Analyzers
	Freezers
	Fertilization tests
	Other appliances for detecting, preventing, monitoring, treating, or alleviating illness, injury, or disability
9. Monitoring and control instruments	Smoke detector
	Heating regulators
	Thermostats
	Measuring, weighing, or adjusting appliances for household or as laboratory equipment
	Other monitoring and control instruments used in industrial installations (e.g., in control panels)
10. Automatic dispensers	Automatic dispensers for hot drinks
	Automatic dispensers for hot or cold bottles or cans
	Automatic dispensers for solid products
	Automatic dispensers for money
	All appliances that deliver automatically all kinds of products

4.1.2 UNITED STATES OF AMERICA

Attention in the United States has been predominantly on electronic waste produced by households (Table 4.2), and as this is considered legally to be a non-hazardous waste (Wagner, 2009), its management and disposal are not dealt with by federal legislation. Consequently, there is no nationwide legislation dealing with WEEE in the United States. Regulations covering such waste are the responsibility of state or municipal authorities. Over the last 10 years, e-waste legislation has been passed in 25 states (2003: California; 2004: Maine; 2005: Maryland; 2006: Washington; 2007: Connecticut, Minnesota, Oregon, Texas, North Carolina; 2008: New Jersey, Oklahoma, Virginia, West Virginia, Missouri, Hawaii, Rhode Island, Illinois, Michigan; 2009: Indiana, Wisconsin; 2010: Vermont, South Carolina, New York, Pennsylvania; 2011: Utah) (http://www.epa.gov/epawaste/conserve/materials/ecycling/index.htm; http://www.epa.gov/epawaste/conserve/materials/ecycling/docs/fullbaselinereport2011.pdf).

4.1.3 INDIA

The Indian government's legislation on WEEE came into force on May 1, 2012. In concept, it is similar to the EU directives, with responsibilities being defined for production, consumption, collection, dismantling, and recycling of e-waste. Two categories of electrical and electronic equipment are defined:

* information technology and electronic equipment
* consumer electrical and electronics

TABLE 4.2
Categories of WEEE Defined by USEPA

Computers	Desktop central processing units (CPUs) and portables
	Computer displays (CRT and flat-screen monitors)
	Keyboards and mice
Televisions	CRTs
	Flat-panel
	Projection
	Monochrome
Hard copy devices	Printers
	Fax machines
	Scanners
	Copiers
	Multi-function devices
Mobile devices	Cell phones
	Smart phones
	Personal digital assistants (PDAs)
	Pagers

The section dealing with reduction in the use of hazardous substances is very similar to the EU RoHS Directive (2002/95/EC), putting restrictions on the use of the same group of contaminants: cadmium, hexavalent chromium, lead, and mercury, and polybrominated biphenyls (PBBs) and polybrominated diphenyl ether (PBDE) flame-retardants (http://www.moef.nic.in/downloads/rules-and-regulations/1035e_eng.pdf).

4.1.4 CHINA

The development of WEEE legislation in China has been well documented by Chung and Zhang (2011). Again, the EU model has been followed and Chinese regulations are covered in three key documents:

- *Management Measure for the Prevention of Pollution from Electronic Information Products (2006)*, which has been called China RoHS, and came into effect on March 1, 2007. Similar in design and intention to EU Directive 2002/95/EC, it has three main features: pollution reduction can be achieved through "design for environment"; the same group of substances defined in the EU Directive are considered hazardous; and it introduces the concept of an "environmental expiry date," which is intended to define the lifetime of products during which hazardous substances in the product will become a health or environmental hazard.
- *Management Measure for the Prevention of Environmental Pollution from Electrical and Electronic Waste*, which came into force in February 2008. This puts a responsibility on WEEE recycling and treatment facilities to carry out an Environmental Impact Assessment.
- *Management Regulations on the Recycling and Disposal of Waste Electrical and Electronic Equipment*, which came into force on January 1, 2011. This document defines specific classes of WEEE: television sets, refrigerators and freezers, washing machines, air conditioners, and computers.

4.1.5 DEVELOPING COUNTRIES

The situation in developing countries is only now being realized, and in particular the effect of imports of old equipment. Very often control on WEEE in these countries depends on the Basel Convention on the Control of Transboundary Movements of Hazardous Wastes and their Disposal, which has been in force since 1992, and which particularly tries to address the recycling, recovery, and disposal of WEEE. There is considerable concern in relation to African countries (Secretariat of the Basel Convention, December 2011, http://www.basel.int/). In Africa specifically, the Bamako Convention on the Ban on the Import into Africa and the Control of Transboundary Movement and Management of Hazardous Wastes within Africa has been in force since 1998. In Asia, too, many countries import WEEE, even when there is legislation in place to ban this, or to allow it only under controlled conditions.

Of the Latin American countries, Argentina, Colombia, and Costa Rica have some limited WEEE legislation, but no country has legislation dealing with WEEE in a comprehensive way (de Oliveira et al., 2012)

4.2 PRODUCTION OF WEEE

Unambiguous statistics for the production of WEEE are difficult to find due to variations in reporting. Robinson (2009) discussed the various statistics and estimated annual global e-waste production of 20 to 25 million tonnes.

Figures from the UK Environment Agency show a total of 517,138.122 tonnes of household (499,022.939 t) and non-household (18,115.183 t) WEEE collected by Producer Compliance Schemes and their members in the UK in 2011 (http://www. environment-agency.gov.uk/business/topics/waste/111016.aspx). This waste is categorized into the 10 EU categories (Table 4.1), with three additional categories separated out: Display Equipment, Cooling Appliances Containing Refrigerants, and Gas Discharge Lamps. Most of the waste (47.37%) falls into the Large Household Appliances and Cooling Appliances Containing Refrigerants categories, with 33.75% in the IT and Telecommunications Equipment, Consumer Equipment and Photovoltaic Panels, and Display Equipment categories. In 2003, it was reported that 940,000 tonnes of domestic WEEE were discarded in the UK (Dalrymple et al., 2007).

Ongondo et al. (2011), collating figures for the EU between 2000 and 2008, suggested that nearly half of the WEEE collected in the EU was Large Household Appliances (49.07%), approximately one-fifth was Consumer Equipment (21.10%), and one-sixth was IT and Telecommunications Equipment (16.27%) (see Table 4.1 for categories). Goosey (2004) quoted a figure of 6.5 million tonnes as the annual amount of WEEE in Europe, estimated to rise to 12 million tonnes per year by 2015. More recently, WRAP (2012) quoted an annual rate of WEEE production in the EU of around 9 Mt, estimated to rise to 10.2 Mt by 2014, and 12.3 Mt by 2020.

In the United States, the Environmental Protection Agency Office of Resource Conservation and Recovery estimated that, in 2009, 2.37 million short tons of electronic products were ready for end-of-life management, but that only 25% of these were collected for recycling. A particular problem in the United States is storage of old electronic equipment by individuals, and the USEPA estimated that there were 5 million short tons in storage in 2009 (http://www.epa.gov/waste/conserve/materials/ecycling/manage.htm).

Countries such as China and India are seeing a rapid increase in the use of electrical and electronic goods, and hence in the production of WEEE. In 2006, Li et al. (2006) estimated that by 2010 there would be approximately 58 M TV sets, >9 M refrigerators, >11 M washing machines, >12 M air conditioners and >70 M PCs obsolete in China. As in the United States, a major problem is that people store these items and do not dispose of them. UNEP (2009), collating figures from a number of sources, gave estimates of approximately 440,000 tonnes of WEEE generated per year in India, and approximately 2,200,000 tonnes of WEEE generated per year in China.

All of these figures must be viewed with a degree of caution because of the different methods of reporting and the different timescales used. For example,

Ongondo et al. (2011) used the 2003 value for WEEE generation in the UK of 940,000 tonnes from Dalrymple et al. (2007), whereas the figure reported for 2011 by the UK Environment Agency was 520,000 tonnes. The former value was esti-mated for "domestic equipment discarded," whereas the latter figure was reported by "Producer Compliance Schemes and their members." The values for WEEE produced in South American countries quoted by de Oliveira et al. (2012) were taken from 2010 figures in http://www.residuoselectronicos.net/?p=2408 (RELAC, Plataforma Regional de Residuos Electrónicos en Latinoamérica y el Caribe, Panorama de RAEE en Latinoamerica). The values of 300,000 t/y for Mexico, 110,000 t/y for Colombia, and 100,000 t/y for Peru compare with 270,000, 36,000, and 24,000, respectively, quoted by UNEP (2009).

An alternative way of viewing the WEEE production estimates is as kilogram pro-duced per person per year (Table 4.3). The target for per capita WEEE collection within the EU is 4 kg/person/year. Statistics published by the EU (http://epp.eurostat.ec.europa.eu/portal/page/portal/waste/data/wastestreams/weee) show WEEE collection figures for 2008 across 28 countries ranging from 0 to 14.8 kg/person/year, with a mean of 5.7.

4.3 COMPOSITION OF WEEE AND POTENTIAL FOR RECOVERY AND RECYCLING

Precise figures for the composition of WEEE are difficult to quantify due to the wide range of products that feed into the waste stream (Table 4.1). The values given by Widmer et al. (2005) are the basis of most estimates (e.g., Ongondo et al., 2011). A very approximate breakdown would be: ~50% ferrous metals; ~20% plastics; ~13% non-ferrous metals; and ~17% other materials such as printed circuit boards, cables, CRT and LCD screens, and pollutants.

Most legislation dealing with WEEE relies on collection and take-back strategies. Much of this is based on the principle of extended producer responsibility (EPR), which has two main goals:

- to put on to manufacturers responsibility for all aspects of a product through-out its lifecycle, but especially for take-back, recovery, and disposal;
- development of more sustainable products, including resource recovery from end-of-life products.

In economically developed countries, such strategies work by and large, but there are still pressures on achieving recovery targets resulting from inappropri-ate disposal or storage of obsolete equipment. The main concern in these countries is contamination of the environment. In less-developed countries, enforcement of legislation and availability of the necessary infrastructure are often impediments to dealing with WEEE. As a result, there is a high rate of repair and reuse in these countries, and often informal recycling using dangerous practices. The main concern in less-developed countries is therefore that of human health due to exposure to toxins contained in the waste.

A further driving force for promotion of recovery and recycling of materials from WEEE is the question of resource depletion and sustainability. The recovery of

TABLE 4.3
Estimates of WEEE Production

Country	WEEE Produced in tonnes Per Year (Rounded figures)	Year of Data	WEEE Generated (kg/person)
Europe			
EU[a]	9,500,000	2008	
Germany[b]	1,100,000	2005	13.3
United Kingdom[c]	520,000	2011	15.8
Switzerland[b]	66,000	2003	9.0
North and South America			
USA[b]	2,250,000	2007	7.5
Brazil[d]	370,000	2006	3.5
Mexico[e]	300,000	2010	2.7
Canada[b]	86,000	2002	2.7
Colombia[e]	110,000	2010	2.4
Peru[e]	100,000	2010	3.4
Asia			
China[d]	2,200,000	2007	1.7
Japan[b]	860,000	2005	6.7
India[d]	440,000	2007	0.4
Africa			
Nigeria[f]	1,000,000	2010	7.1
Ghana[f]	179,000	2009	7.5
South Africa[d]	60,000	2007	1.2
Morocco[d]	38,000	2007	1.4
Côte d'Ivoire[f]	15,000	2009	0.7
Benin[f]	9700	2009	11.1
Kenya[d]	7000	2007	0.2
Senegal[d]	4000	2007	4.9
Uganda[d]	2500	2007	0.1

[a] WRAP (2012)
[b] Ongondo et al. (2011)
[c] Environment Agency
[d] UNEP (2009)
[e] de Oliveira et al. (2012)
[f] Secretariat of Basel Convention (2011)

metals, for example, could reduce the need for further mining, with the advantage of retaining scarce resources. There would also be other benefits to the environment, such as lower energy use and lower CO_2 emissions.

There is particular concern about the future supply of certain critical elements needed for EEE production (WRAP, 2012). In 2010, the EU identified 14 critical raw materials: antimony, beryllium, cobalt, fluorspar, gallium, germanium, graphite, indium, magnesium, niobium, platinum group elements (PGEs), rare earth elements (REEs), tantalum, and tungsten. Table 4.4 summarizes the production and demand for the common metals used in production of EEE. Although these are average values based on a number of assumptions, it would seem that recovery from WEEE could supply approximately 30% of the annual production for silver and the plantinum group elements (especially palladium); and for just under 20% of the demand for antimony, cadmium, gold, and tin. Looked at in terms of meeting annual demand requirements, recovery from WEEE could supply all the need for gold, PGEs, and silver, and about half the annual demand for antimony and tin. Based on the global sales in 2006, Hagelüken (2008) estimated that recovery from mobile phones, PCs, and laptops alone could account for 2.5% of the annual mine production of silver, 3% for gold, 12% for palladium, and 1% for copper. Recovery from the batteries in these products could account for 15% of cobalt production.

UNEP (2009), using data from the EcoInvent 2.0 database, calculated that the mining of copper used for EEE production (based on 2006 figures) generated 15.30 Mt CO_2; other significant emissions were generated by mining of gold (5.10 Mt CO_2), tin (1.45 Mt CO_2), and silver (0.86 Mt CO_2). Given the estimates shown above for the amount of production or demand of metals for use in EEE that could be met by recycling from WEEE, there is clearly considerable potential for reductions in CO_2 emissions associated with these products.

UNEP (2009) also calculated that 1 kg aluminium produced by recycling of WEEE would require less than 10% of the energy used to produce that aluminium by smelting bauxite. In addition, 1.3 kg red mud, 2 kg CO_2 emissions, and 0.11 kg SO_2 emissions would be avoided. Therefore, there would be environmental benefits arising from less energy use, lower emissions, and less solid waste that would require to be stored.

Cui and Forssberg (2003) estimated the savings in energy obtained from the use of recycled as opposed to mined materials as: aluminium, 95%; copper, 85%; plastics, >80%; iron and steel, 74%; lead, 65%; paper, 64%; and zinc, 60%.

4.4   RECYCLING PROCESSES

The variety of products that contribute to the WEEE stream is extremely diverse—a mobile phone is very different from a refrigerator—and the materials that can be recovered also differ. However, there are some steps that are common to most treatments of WEEE (Figure 4.1).

The collection and initial dismantling are usually done on a local basis. Recovery of iron/steel and aluminium also tend to be done in smelters close to points of collection. The more specialized recovery of precious metals and batteries is focused in a few dedicated facilities in Belgium, Canada, Germany, Japan, and Sweden (UNEP, 2009).

A detailed discussion of recycling and recovery practices worldwide is given in WRAP (2012), and for the Umicore Integrated Metals Smelter and Refinery near Antwerp, Belgium, in Hagelüken (2006).

TABLE 4.4

Estimates of Production and Demand of Some Critical Metal Resources Used in EEE

Metal	Annual mine production (t/yr)[a]	Estimated reserves (tonnes)[b]	Reserves/ annual mine production	Demand for EEE (t/yr)[c]	Demand as % of production	Typical concentration in WEEE (mg/kg)[d]	Annual global amount of metals in WEEE (tonnes)[e]	Potential recovery as % of demand	Potential recovery as % of production
Antimony (Sb)	175,000	1,800,000	10	65,000	37	1700	34,000	52	19
Cadmium (Cd)	20,000	640,000	32	—		180	3600		18
Chromium (Cr)	22,000,000	>480,000,000	22	—		9900	198,000		0.9
Cobalt (Co)	70,000	7,500,000	107	11,000	16				
Copper (Cu)	15,400,000	690,000,000	45	4,500,000	29	41,000	820,000	18	5
Gold (Au)	2455	51,000	21	300	12	22	440	147	18
Lead (Pb)	3,800,000	85,000,000	22	—		2900	58,000		1.5
Mercury (Hg)	1600	93,000	58	—		0.68	13.6		0.85
Nickel (Ni)	1,565,000	80,000,000	51			10,300	206,000		13
PGEs	460	66,000	143	46	10	7	140	98	30
Silver (Ag)	21,560	530,000	25	6,000	28	313	6260	104	29
Tin (Sn)	284,000	4,800,000	17	90,000	32	2400	48,000	53	17
Zinc (Zn)	1,110,000	250,000,000	225	—		5100	102,000		9

[a] Rounded figure from data for 2004–2008 from BGS (2010) and 2010 and 2011 from USGS (2012).

[b] USGS (2012).

[c] UNEP (2009) based on demand in 2006.

[d] Values from Morf et al. (2007); Bigum et al. (2012).

[e] Assuming WEEE production of 20 Mt per year (Robinson, 2009).

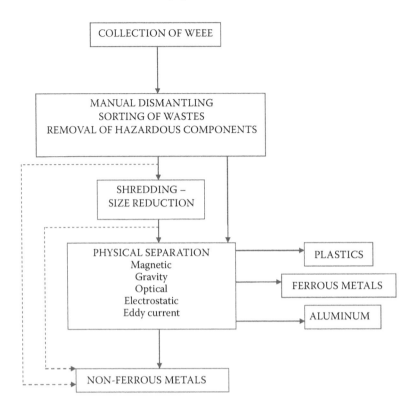

FIGURE 4.1   Outline of processes used for recycling WEEE.

4.4.1   COLLECTION OF WEEE

A crucial step in the overall process of recycling is the collection of WEEE, which, as has been described previously, is the focus of much of the legislation that has been put in place across the world (collection centers, manufacturer takeback schemes, etc.). Despite these measures, the efficiency of collection rates varies widely. Khetriwal et al. (2011) gave average collection rates within the EU of 40% for large appliances and 25% for medium-sized appliances, with small appliances being the greatest problem with some countries having collection rates of about zero. There may be a number of reasons for these poor collection rates: lack of user awareness about recycling, lack of awareness among retail staff regarding takeback schemes, lack of collection infrastructure, storage of obsolete equipment by consumers, cost, and illegal exporting of WEEE.

4.4.2   MANUAL DISMANTLING/SORTING/REMOVAL OF HAZARDOUS COMPONENTS

Although automated disassembly would be desirable, manual operations have proven to be more efficient for two reasons (Opalic et al., 2010):

- the high variety of products classified as WEEE
- unfavorable design

This stage of the process has two key functions:

- separating components efficiently so that they can be directed toward the appropriate further treatment process
- safe removal and treatment of toxic or environmentally damaging materials

The balance between these two functions depends on the type and age of the equipment. Large appliances such as washing machines and refrigerators would at one time have contributed waste mostly into the ferrous metal stream, with smaller amounts of plastic, glass, etc. More recently, such appliances may also have circuit boards as electronic controls have been introduced. Refrigerators and air conditioning units require their refrigerants to be removed. In older models, these may have been chlorofluorocarbons (CFCs) or hydrochlorofluorocarbons (HCFCs), which are potent greenhouse gases and cause depletion of the ozone layer. Small equipment, such as mobile phones, is more important as a source of other materials. For example, Hagelüken (2008) estimated the average composition of 150 tonnes of mobile phone handsets recovered at the Unicore treatment facility in Belgium to be 43% organics, 34% "others" (mainly glass and ceramics), 13% copper, 7% iron, 1.4% nickel, 1.1% zinc, 1% tin, 0.6% lead, and 0.4% precious metals (3512 g silver/tonne, 341 g gold/tonne, 144 g palladium/tonne, 4 g platinum/tonne).

Cui and Forssberg (2003) listed the main hazards found in WEEE:

- Batteries (heavy metals)
- Cathode ray tubes (lead)
- Liquid crystal displays (mercury)
- Gas discharge lamps (mercury)
- Mercury-containing components (switches, thermostats, sensors, relays)
- Printed circuit boards (heavy metals)
- Polychlorinated biphenyls (PCBs) (capacitors)
- Toner cartridges
- Plastics with halogenated flame retardants
- Fluorocarbon refrigerants
- Asbestos

The type of hazard may also change with newer technology, for example, with display equipment. The older cathode ray tubes (CRTs) were a major source of lead, while the more recent liquid crystal displays (LCDs) contain mercury. Many of these hazards are also potential resources (e.g., batteries and printed circuit boards), while others are simply hazards that must be disposed of safely (e.g., asbestos).

An important aspect of this stage of the recycling process is the efficient separation of materials. This is important for two reasons:

- to ensure minimum loss of recovered resource, which could occur if materials are forwarded to the wrong treatment process (e.g., precious metals would not be recovered in the ferrous metal or aluminium streams).

- to ensure that hazards are not created at the next stage by the presence of materials that could produce contaminants (e.g., organic components in the ferrous metal stream could produce volatile organic compounds, dioxins, and furans during the smelting process).

UNEP (2009) identified three levels of potential toxic emissions from treatment of WEEE:

- Primary emissions: toxic components of the WEEE (heavy metals, PCBs, fluorocarbons).
- Secondary emissions: products of inappropriate treatment (dioxins produced by incineration of organics).
- Tertiary emissions: toxic materials used at later stages of the recycling process (cyanide, mercury).

This stage may be preceded or followed by crushing, breakdown, or shredding of the WEEE. In some cases, this may ease the removal of components such as printed circuit boards, and so would be done prior to dismantling, while in other cases removal of valuable or hazardous components may be complete and just the structural units are broken down.

4.4.3 Resource Recovery

4.4.3.1 Ferrous Metals

These contribute typically approximately 50% of a WEEE stream. Very largely these come from metal casings for equipment such as washing machines, refrigerators, air conditioning units, etc., which can be readily separated by magnetic means. These materials are used in metal smelters to produce more iron and steel. There are issues regarding paint and other coverings, which may affect the smelting process, and which could produce toxic gaseous emissions. It is also important to effectively remove other metals, which could affect the remelting of the steel, and to maximize their recovery. While shredding and magnetic or manual separation can be effective in some cases (UNEP, 2009), Chancerel et al. (2008) estimated that, in the process they studied, 6% of the copper, 35% of the silver, 40% of the gold, and 23% of the palladium ended up in the ferrous metals fraction following separation.

4.4.3.2 Aluminium

Morf et al. (2007) reported a typical value of approximately 5% aluminium in small WEEE from consumer and commercial equipment in Switzerland, and Bigum et al. (2012) quoted a value of 2 to 3% aluminium in "high grade" WEEE.

Aluminium can be recovered from WEEE by heating to 700 to 800°C, but this represents a small fraction of the total scrap aluminium available with drinks cans, food packaging, and structural materials feeding in far more to the waste stream. The big attraction to recycling aluminium is the savings of approximately 90% in energy costs compared to smelting of bauxite. Again, there are problems of paint and other coatings, and of the presence of other metals in the scrap. The WEEE aluminium can be mixed with scrap from other sources in order to obtain the required alloy composition. Melting is carried out with a flux of sodium and potassium chloride, which removes impurities and helps to prevent oxidation of the aluminium, although some will occur. The salt with its impurities is separated from the molten aluminium, and can be a significant amount—up to 500 kg per tonne of aluminium. This salt slag can be further treated to recover aluminium metal and aluminium oxide, and the dissolved salts can be recrystallized and reused.

4.4.3.3   Precious Metals

Because of the value of the metals, this group of materials is the one that has received the greatest research effort on the recycling of WEEE. The metals listed in Table 4.4 are the ones usually considered under this heading, but there has been particular interest in copper, gold, silver, and the platinum group elements (especially palladium).

Cui and Zhang (2008) and Tuncuk et al. (2012) have reviewed the methods used to recover metals from WEEE, which fall into four broad categories:

- Pyrometallurgical processes—smelting, melting, incineration, drossing, sintering, and gas phase reactions at high temperature.
- Hydrometallurgical processes—leaching with solutions of specific leaching agents (acid, alkali, halides, cyanide, thiosulfate, thiourea), followed by separation and purification steps (precipitation, solvent extraction, adsorption, ion exchange).
- Biometallurgical processes—bioleaching, where metal removal is enhanced by microbial action; biosorption, physico-chemical interactions between metal ions in solution and the surfaces of living or dead microorganisms.
- Electrometallurgical processes—electro-winning/refining.

The application of these techniques is described by Hagelüken (2006) at the Umicore Integrated Metals Smelter and Refinery at Hoboken near Antwerp in Belgium.

4.4.3.4   Plastics

Table 4.5 shows the main plastics used in the manufacture of EEE, and examples of their uses. The most widely used are HIPS, ABS, PC, and various forms of PE, and these constitute approximately 20% on a weight basis in a typical WEEE stream. These materials are usually produced as shredded material following removal of metals and hazardous components. There are two impediments to effective recycling of these plastics (Schlummer et al., 2006):

- the complex mixture of different types (Table 4.5).
- the presence of toxic materials—especially polybrominated biphenyls (PBBs) and polybrominated diphenyl ether (PBDE) used as flame retardants, which may form brominated furans and dioxins when subjected to heat.

As these factors have been recognized, there has been a drive to reduce the range of plastics used in EEE; and as the use of PBBs and PBDEs has been restricted by the

TABLE 4.5
Plastics Used in Production of EEE

Type of Plastic	Abbreviation	Properties	Examples of Products
Styrenics			
Polystyrene	PS		Refrigerator trays/ linings, TV cabinets
High-impact polystyrene	HIPS	Good impact protection	
Acrylester-styrene-acrylonitrile	ASA		
Styrene-acrylonitrile	SAN		Hi-fi covers
Acrylonitrile-butadiene-styrene	ABS	Good impact protection; chemical and heat resistant (Achilias et al., 2009)	Telephone handsets, keyboards, monitors, computer housings
Non-Styrenics			
Polypropylene	PP		Kettles
Polyethylene	PE	Good electrical insulation	Cable and wire insulation
Polyphenylene ether	PPE		
Polyphenylene oxide	PPO	High temperature resistance, rigidity, high impact protection	Coffee machines, TV housings
Polycarbonate	PC	Transparency, good thermal stability, high heat distortion temperature, flame retardency (Achilias et al., 2009)	Telephones
Polyurethane	PU		
Polyamide	PA		Food processor bearings, adaptors
Polyvinyl chloride	PVC	Good electrical insulation	Cable and wire insulation, cable trunking
Blends of polycarbonate/ Acrylonitrile-butadiene-styrene	PC/ABS		
Blends of high-impact polystyrene/poly (p-phenylene oxide)	HIPS/PPO		

RoHS Directive (2002/95/EC), alternative flame-retardants or plastics have been used. In the past, the high cost of recycling plastics from WEEE meant that much of it was disposed of to landfill. More recently, however, more cost-effective techniques have been developed (Cui and Forssberg, 2003; Schlummer et al., 2006; Achilias et al., 2009).

There are broadly two approaches taken to recycling plastic from WEEE:

- dissolution/reprecipitation, which attempts to recycle polymers.
- pyrolysis, which attempts to produce economically valuable chemicals by degradation of plastics.

The dissolution/reprecipitation approach has the drawback that it requires the use of large quantities of solvents, some of which may be toxic, and that it may be applicable only to certain classes of plastics (Vilaplana and Karlsson, 2008). Ideally, the solvent used should have a high ability to dissolve the target polymer with minimum degradation of the polymer chain.

Achilias et al. (2009) used dichloromethane, acetone, toluene, or chloroform in a 2:3 mix with methanol to extract ABS, PC, or PS from a range of plastic parts separated from commercial WEEE, and compared with extraction of the model polymer. They reported high recovery rates of pure polymer from the WEEE components (Table 4.6).

Schlummer et al. (2006) identified the styrene group of polymers (Table 4.5) as having particularly high value and so being worth recovering from WEEE. However, there are incompatibilities within this group that could lead to problems of decreased impact strength or ability to withstand strains. Thus, any recycling process should remove the incompatible polymers or, alternatively, additives to improve the property of the recycled polymer must be added. Schlummer et al. (2006) described a two-step process for the separation of styrenic polymers (HIPS and ABS) based on density separation followed by the CreaSolv® solvolysis, which removes unwanted polymers and brominated flame retardants, furans, and dioxins. They reported 50% recovery at the first stage and 70 to 80% recovery in the second stage.

The alternative strategy for recycling plastics from WEEE is pyrolysis, which involves heating the waste to moderate temperatures in the absence of oxygen

---

TABLE 4.6
Extraction of Polymers from WEEE

WEEE Type	Polymer	Solvent Mix[a]	Temperature (°C)	Recovery (wt%)
Computer monitor	ABS	Dichloromethane	100	91
		Toluene	100	70
		Acetone	25	68
TV set	ABS	Dichloromethane	100	96
		Toluene	100	69
		Acetone	25	90
Electronic toy	ABS	Dichloromethane	100	90
		Toluene	100	61
CD	PC	Dichloromethane	50	98
Radio	PS	Dichloromethane	100	98
		Toluene	100	95
Vacuum cleaner	PS	Dichloromethane	100	95
		Toluene	100	91
Electronic toy	PS	Dichloromethane	100	95
		Toluene	100	97

Source: Achilias et al., 2009.

[a] All in a 2:3 mix with methanol.

TABLE 4.7

Elemental Composition and Distribution of Pyrolysis Products from WEEE

Type of WEEE	C (wt%)	N (wt%)	H (wt%)	Other (wt%)	Ash (wt%)	Liquid	Gas	Solid[a]	Char
Hall and Williams (2007)									
CRT	81.6	5.5	7.5	4.2	1.3	83.9	1.4	14.5	13.2
Refrigeration	71.4	1.8	7.0	6.0	13.8	76.5	3.0	20.4	6.6
Mixed WEEE	75.7	0.8	6.3	13.8	3.3	70.6	7.8	21.1	17.8
de Marco et al. (2008)									
PE wires	64.1	0.0	9.6	—	28.0	44.1	23.0	32.9	4.9
Table phones	75.1	4.8	6.6	—	15.0	53.5	12.2	34.4	19.4
Mobile phones	70.1	1.8	5.7	6.8	15.6	57.4	12.3	30.3	14.7
Printed circuit boards	20.5	0.5	1.4	6.9	70.7	16.2	7.3	76.5	5.8

[a] Solid = ash + char

(de Marco et al., 2008). The plastics are decomposed to gaseous or liquid products, which can be isolated, whereas the inorganic components (e.g., contaminant metals) are unchanged. There is also a char formed during the pyrolysis process. Two recent studies have reported the results of pyrolysis of WEEE-derived materials. Hall and Williams (2007) treated three types of WEEE (plastics from CRT-containing equipment [TV sets, computer monitors], shredded plastic from refrigeration equipment, and mixed WEEE plastics) in a fixed bed reactor under nitrogen at 600°C. de Marco et al. (2008) treated four types of WEEE (polyethylene wire coatings, shredded table phone casings, shredded mobile phone casings, and shredded printed circuit boards) in an unstirred stainless steel autoclave under nitrogen at 500°C. The elemental composition and the distribution of pyrolysis products are given in Table 4.7. In both studies, the main gases produced were CO, CO_2, H_2, and C_1–C_5 hydrocarbons. The liquids obtained in each case were a complex mix—mainly aromatic and nitrogenated compounds. While these were both initial studies, they showed particular potential for producing styrene and phenol, but much more work is needed on isolation and separation of individual products.

Use of plastic in WEEE as an energy source has also been suggested using the plastic material itself (Menad et al., 1998; Vehlow et al., 2000) or the gases produced by pyrolysis (de Marco et al., 2008). The main concern about using plastics, even when mixed with other fuels, is the production of toxic compounds. Of particular concern is the production of dioxins and furans due to incomplete combustion; this is exacerbated in the case of WEEE plastics due to the potential presence of brominated flame retardants producing brominated compounds. There is also concern about production of other halogenated compounds, from combustion of PVC for example.

Another potential use for waste plastics is as a replacement for aggregate in concrete (Ismail and Al-Hashmi, 2008; Bagel and Matiasovsky, 2010).

REFERENCES

Achilias, D.S., E.V. Antonakou, E. Koutsokosta, and A.A. Lappas. 2009. Chemical recycling of polymers from waste electric and electronic equipment. *Journal of Applied Polymer Science* 114: 212–221.

Bagel, L., and P. Matiasovsky. 2010. Surface pretreatment—a way to effective utilization of waste plastics as concrete aggregate. Review and first experiences. In: *Cesb 10: Central Europe Towards Sustainable Building—from Theory to Practice*, P. Hajek, J. Tywoniak, A. Lupisek, J. Ruzicka, and K. Sojkova, Eds. Czech Technical University, Prague, pp. 355–358.

Bigum, M., L. Brogaard, and T.H. Christensen. 2012. Metal recovery from high-grade WEEE: a life cycle assessment. *Journal of Hazardous Materials* 207–208: 8–14.

Chancerel, P., C. Meskers, C. Hagelüken, and S. Rotter. 2008. E-scrap metals too precious to ignore. *E-Scrap Research*, November 2008.

Chung, S.-S., and C. Zhang. 2011. An evaluation of legislative measures on electrical and electronic waste in the People's Republic of China. *Waste Management* 31: 2638–2646.

Cui, J., and E. Forssberg. 2003. Mechanical recycling of waste electric and electronic equipment: a review. *Journal of Hazardous Materials* 99: 243–263.

Cui, J., and L. Zhang. 2008. Metallurgical recovery of metals from electronic waste: a review. *Journal of Hazardous Materials* 158: 228–256.

Dalrymple, I., N. Wright, R. Kellner et al. 2007. An integrated approach to electronic waste (WEEE) recycling. *Circuit World* 33: 52–58.

de Marco, I., B.M. Caballero, M.J. Chomón et al. 2008. Pyrolysis of electrical and electronic wastes. *Journal of Analytical and Applied Pyrolysis* 82: 179–183.

de Oliveira, C.R., A.M. Bernardes, and A.E. Gerbase. 2012. Collection and recycling of electronic scrap: A worldwide overview and comparison with the Brazilian situation. *Waste Management* 32: 1592–1610.

Goosey, M. 2004. End-of-life electronics legislation—an industry perspective. *Circuit World* 30: 41–45.

Hagelüken, C. 2006. Recycling of electronic scrap at Unicore's integrated metals smelter and refinery. *World of Metallurgy—ERZMETALL* 59: 152–161.

Hagelüken, C. 2008. Opportunities & challenges to recover scare and valuable metals from electronic devices. Paper presented to the OCED-UNEP Conference on Resource Efficiency, Paris, April 24, 2008.

Hall, W.J., and P.T. Williams. 2007. Analysis of products from the pyrolysis of plastics recovered from the commercial scale recycling of waste electrical and electronic equipment. *Journal of Analytical and Applied Pyrolysis* 79: 375–386.

Ismail, Z. Z., and E.A. Al-Hashmi. 2008. Use of waste plastic in concrete mixture as aggregate replacement. *Waste Management* 28: 2041–2047.

Khetriwal, D.S., R. Widmer, R. Kuehr, and J. Huisman. 2011. One WEEE, many species: lessons from the European experience. *Waste Management & Research* 29: 954–962.

Li, J., B. Tian, T. Liu, H, Liu, X. Wen, and S. Honda. 2006. Status quo of e-waste management in mainland China. *Journal of Material Cycles and Waste Management* 8: 13–20.

Menad, N., Bo. Björkman, and E.G. Allain. 1998. Combustion of plastics contained in electric and electronic scrap. *Resources, Conservation and Recycling* 24: 65–85.

Morf, L.S., J. Tremp, R. Gloor, F. Schuppisser, M. Stengle, and R. Taverna. 2007. Metals, non-metals and PCB in electrical and electronic waste Actual levels in Switzerland. *Waste Management* 27: 1306–1316.

Ongondo, F.O., I.D. Williams, and T.J. Cherrett. 2011. How are WEEE doing? A global review of the management of electrical and electronic wastes. *Waste Management* 31: 714–730.

Opalic, M., M. Kljajin, and K. Vučkovič. 2010. Disassembly layout in WEEE recycling process. *Strojarstvo* 52: 51–58.

Robinson, B.H. 2009. E-waste: An assessment of global production and environmental impacts. *Science of the Total Environment* 408: 183–191.

Schlummer, M., A. Maurer, T. Leitner, and W. Spruzina. 2006. Report: Recycling of flame-retarded plastics from waste electric and electronic equipment (WEEE). *Waste Management & Research* 24: 573–583.

Secretariat of the Basel Convention. 2011. Where are WEee in Africa? Findings from the Basel Convention E-Waste Africa Programme.

Tuncuk, A., V. Stazi, A. Akcil, E.Y. Yazici, and H. Deveci. 2012. Aqueous metal recovery techniques from e-scrap: Hydrometallurgy in recycling. *Minerals Engineering* 25: 28–37.

UNEP. 2009. Recycling—From E-waste to Resources.

Vehlow, J., B. Bergfeldt, K. Jay, H. Seifert, and T. Wanke. 2000. Thermal treatment of electrical and electronic waste plastics. *Waste Management & Research* 18: 131–140.

Vilaplana, F., and S. Karlsson. 2008. Quality concepts for the improved use of recycled polymeric materials: a review. *Macromolecular Materials and Engineering,* 293: 274–297.

Wagner, T.P. 2009. Shared responsibility for managing electronic waste: a case study of Maine, USA. *Waste Management* 29: 3014–3021.

Widmer, R., H. Oswald-Krapf, D. Sinha-Khetriwal, M. Schnellmann, and H. Bőni. 2005. Global perspectives on e-waste. *Environmental Impact Assessment Review* 25: 436–458.

WRAP. 2012. Strategic Raw Materials, Recovery Capacity and Technologies.

5 Food Waste Utilization

N. Dunn

CONTENTS

5.1 INTRODUCTION

Globally, it is estimated that approximately 1.3 billion tonnes per year of the edible parts of food produced for human consumption is wasted (Gustavsson et al., 2011). Therefore, food waste has significant potential as a valuable resource to be managed more efficiently. The primary aim must be to reduce the quantity of waste produced in the first instance; however, effective resource management when food becomes waste can also be useful for extracting the highest value from the material.

Food waste production is recognized as a significant issue in developed countries. In the most affluent societies, consumers waste large quantities of food. It has been estimated that approximately 89 million tonnes of food waste are generated in the EU per year (Eurostat, 2011). An estimated 36 million tonnes of municipal food waste was generated in the United States in 2011 (USEPA, 2010) and it is estimated that the amount has increased by 50% since 1974 (Pittman, 2013). In Australia, total food waste (household and commercial) is estimated at 7.57 million tonnes per year (Australian Government, 2010).

There are concerns that as BRIC (Brazil, Russia, India, and China) countries develop, they will also begin to waste food at the high levels seen in OECD countries. Already, it has been estimated by experts that 26 million tonnes of food are wasted in Brazil each year (Government Office for Science, 2011).

Food waste is a problem in developing countries, too, but there are significant differences in the ways in which food is wasted when compared to the developed world. In the developing world, most food waste is considered to be caused by lack of appropriate storage and inadequate distribution systems rather than as a result of consumer wastage (Dorward, 2012). Okonko et al. (2009) gave an overview of the utilization of wastes (including food wastes) from an African perspective. The authors consider that developing countries are now dealing with the economic growth and developmental issues faced by Western economies in the 1970s while also dealing with their own socio-economic issues.

Pre-consumer waste can also occur in developed countries, but for different reasons such as buyers' aesthetic requirements, over-production as insurance against a poor harvest that might lead to under-supplying contracts, standardized manufacturing processes, and a reliance on cautious "sell by" dates (Dorward, 2012).

Making the most efficient use of food as a resource is therefore of global importance. This chapter will provide examples of the policy landscape, and consider the potential options for food waste utilization to make the best use of the available resource.

5.2   FOOD WASTE POLICY

Food waste is a growing policy concern due to a growing world population putting increased pressure on limited raw resources. Difficult economic conditions provide reasons to maximize resource efficiency and reduce the costs associated with waste. The environmental costs also keep waste policy on the agenda of many governments.

Tackling the problem of food waste production requires the development of policies to encourage greater resource efficiency and reduce wastage and also to appropriately manage the waste when it is produced.

There is growing interest in resource efficiency and this is enshrined in EU policy through transposition of the revised Waste Framework Directive (WFD) (European Union, 2008) in EU Member States. The revised WFD requires the waste hierarchy to be followed; that is, to reduce, reuse, recycle, and recover waste before disposal options are considered. In the United States, practices that reduce the amount of waste that needs to be disposed of, such as waste prevention, recycling, and composting in line with the waste hierarchy, are encouraged. Similarly, Australia has a National Waste Policy that aims to avoid the generation of waste, reduce the amount of waste (including hazardous waste) for disposal, and manage waste as a resource.

However, despite the aim of these policies to reduce waste production as a priority, there is evidence that the amount of waste being produced is not reducing at present (Von Homeyer et al., 2011; USEPA, 2011), and in some instances waste generation has actually increased. Much of the progress achieved by waste policies is therefore the development of appropriate infrastructure to allow wastes to be treated by recycling or recovery techniques.

5.2.1   DIVERSION FROM LANDFILL

In the EU, a major driver was the introduction of the Landfill Directive (European Union, 1999), which introduced targets for reducing the amount of biodegradable municipal wastes sent to landfills. Landfill diversion targets for each Member State are derived from the quantity of biodegradable municipal wastes landfilled in the reference year (1995). The targets specify a 25% reduction by 2006, 50% reduction by 2009, and 65% reduction by 2016. Member states with a high reliance on landfill as a means to dispose of their waste in 1995 (e.g. UK, Ireland, and Estonia) could apply for a derogation to allow an extension of the target dates by up to 4 years.

The drive to reduce municipal biodegradable waste to landfill has led to increased use of other waste management techniques such as incineration (with and without energy recovery), mechanical biological treatment (MBT), composting, and anaerobic digestion (AD). In the UK, the Landfill Tax escalator has made these alternative options more economically attractive and has been a major policy driver in the shift from landfilling (Defra, 2012).

In the United States, the federal legislation on waste (including municipal, agricultural, and industrial food waste) is the Resource Conservation and Recovery Act (RCRA). The RCRA, which Congress passed in 1976, set national goals for protecting human health and the environment from the potential hazards of waste disposal; conserving energy and natural resources; reducing the amount of waste generated; and ensuring that wastes are managed in an environmentally sound manner. However, the availability of landfills in the United States may account for the failure of widespread uptake of composting biodegradable wastes in many areas (Pittman, 2010).

In South Korea, landfilling of raw food waste has been banned since January 2005 due to a shortage of landfill space, concerns over groundwater and soil contamination, and in order to promote recycling of food as a resource (Kim and Kim, 2010; Behera et al., 2012). The ban meant that 94% of food wastes were recycled in 2006 as animal feed, composting, or by other methods such as AD.

5.2.2 Food Waste as a Resource

Food waste in the developed world therefore represents a significant potential resource, but this depends on several important factors: source and type of waste, availability and proximity to locations where it can be used; cost or incentives; and potential risks.

Many of the uses of food wastes are focused on low value utilization based on the plant or animal nutrient content, or the energy (calorific) content. However, there is growing interest in the biorefining of food waste materials for developing high value products.

5.3 CAUSES AND IMPACTS OF FOOD WASTE

Losses and wastage occur at all parts of the food chain. The Food and Agriculture Organization of the United Nations (FAO) investigated the extent of food waste using five system boundaries: agricultural production, post-harvest handling and storage, processing, distribution, and consumption (Gustavsson et al., 2011). Food losses in industrialized countries are as high as in developing countries, but in developing countries more than 40% of the food losses occur at post-harvest and processing levels, while in industrialized countries, more than 40% of the food losses occur at retail and consumer levels (Gustavsson et al., 2011).

In the UK, the Waste and Resources Action Programme (WRAP) summarized the reasons for food waste production at the household level as: householders buying too much food; shopping without a list; refrigerator temperatures being set too high so food goes off too quickly; and at least 6 out of 10 people throwing food away because it has passed its "use by" date (WRAP, 2007). WRAP estimates that food waste represents approximately 20% of UK domestic waste. There were an estimated 7.2 million tonnes of household food waste in the UK in 2010 (Quested and Parry, 2011).

Outside of the home, food is also wasted for similar reasons by the catering and hospitality industry. WRAP estimates that some 600,000 tonnes of food are disposed of by the hospitality sector (Williams et al., 2011). The Sustainable Restaurant Association conducted research on food waste in restaurants. In the UK reasons for food waste in restaurants were cited as: food is unusable (for example, radish tops, onion skins, banana skins); food is not cooked before spoiling; over-ordering; food is left out too long; incorrect refrigerator temperatures; mistakes in cooking, food falling on floor; over-portioning; and customer food returns (Sustainable Restaurant Association, 2010).

5.3.1 PROBLEMS ASSOCIATED WITH FOOD WASTE

5.3.1.1 Economic Cost

The most obvious impact resulting from any waste is the economic cost associated with its disposal. Landfill taxes are used to create an economic incentive for the development of other waste treatment and disposal methods by making landfill less competitive (Martin and Scott, 2003). High food waste disposal costs make an impact upon the profitability of businesses; increase costs to disposal authorities; and Quested and Johnson (2009) estimated wasted food to cost UK households £480 per year.

5.3.1.2 Global Hunger and Food Security

With a growing global population, there is an ever-increasing demand for food. Limited global resources mean that food security is on the political agenda. However, progress in reducing global hunger has been slowing in recent years (FAO et al., 2012). Reducing hunger and under-nourishment requires action across a number of areas, with reducing food waste and losses noted as an area where gains could easily be made. Gustavsson et al. (2011) note that the issue of food losses is of high importance in the efforts to combat hunger, raise income, and improve food security in the world's poorest countries. Kummu et al. (2012) state that halving food supply losses would provide enough food for approximately 1 billion extra people.

5.3.1.3 Resource Costs

Kummu et al. (2012) investigated the impact that food losses had on other resources, such as freshwater, cropland, and fertilizers. They estimated the resources used to produce lost and wasted food crops from the global food supply. The production of lost and wasted food crops accounts for 24% of total freshwater resources used in food crop production, 23% of total global cropland area, and 23% of total global fertilizer use (Kummu et al., 2012). Reducing food waste would therefore increase the efficiency of resources and inputs used in the production of food.

5.3.1.4 Greenhouse Gas Emissions

Food waste is an important contributor to the global food chain's greenhouse gas emissions. Food waste in landfill decomposes anaerobically, leading to emissions of both carbon dioxide and methane. The manufacture, distribution, storage, use, and disposal of food that is wasted in the UK is associated with greenhouse gas emissions of approximately 17 million tonnes of CO_2 equivalents (Quested and Parry, 2011). The global warming potentials for landfill, incineration, centralized composting, and anaerobic digestion (AD) have been calculated as, respectively, +743, +13, −14, and −170 kg CO_2 equivalents/tonne food waste (Evans, 2012), giving more reason to divert food from landfill as a priority.

5.3.2 RISKS AND CONCERNS

5.3.2.1 Animal By-Products

Animal by-products are entire bodies or parts of animals, products of animal origin, or other products obtained from animals that are not intended for human consumption. Food wastes containing animal by-products may pose a higher risk during treatment or disposal due to the potential for these materials to introduce pathogens into the end product. In the EU, the Animal By-Products Regulations (European Union, 2009) set the conditions that must be met in order for the material to pose a minimal risk when used, treated, or disposed. The regulations do not allow catering wastes to be fed to animals.

5.3.2.2 Plant Material

Food wastes that do not contain animal materials may be considered a lower risk when treated or used. However, there can still be some concerns that wastes may contain harmful bacteria and other pathogens, or residues from plant protection products, the risks of which need to be considered depending on the end-use. Although food industry wastes show potential as substitutes for renewable plant resources used as chemical feedstock, Mahro and Timm (2007) summarized the difficulties when compared to agricultural crops: availability, quality, and price.

5.4 THE WASTE HIERARCHY

The waste hierarchy can be applied to identify the best options for dealing with food waste.

5.4.1 REDUCE

It is most appropriate to find ways to reduce waste production in the first instance, before considering the options for treating or disposing of food as a waste. The options depend very much on the source of food that might be wasted, proximity to other potential users, and the costs involved in diverting the material toward an alternative use. Options include the following.

5.4.1.1 Prevention

Where possible, waste should be prevented from arising. Examples for preventing food waste include businesses undertaking a waste review to understand where their waste arises and what they can do to reduce its occurrence; better portion control in the hospitality industry and households to prevent excessive amounts of food being prepared and wasted; and householders planning their meals to make the best use of the food they buy.

5.4.1.2 Reduction

Where the waste cannot be completely prevented, the amount that is wasted should be reduced. Examples of waste reduction include: businesses donating otherwise edible food to charity, negating the costs of treatment or disposal; supermarkets diverting suitable foods to be re-distributed to charity, re-manufactured into animal feed, or fed directly to animals; farmers ploughing-in spoiled/damaged crops or unharvested parts of plants to make use of nutrients and prevent further disposal costs being incurred; and householders making better use of leftovers to reduce the amount of food being thrown away.

5.4.1.3 Life Cycle Thinking and Waste Prevention

Gentil et al. (2011) used life cycle thinking to model the environmental benefits of waste prevention. Their modeling, based on a fictional European municipality (using average data from central and northern Europe), assessed the benefits of food waste prevention in terms of impacts on a range of environmental criteria including global warming potential, ecotoxicity, nutrient enrichment, human toxicity, and ozone formation. They found that waste prevention could reduce both the net adverse impacts and net benefits of a waste management system. However, the greatest benefit was gained due to avoided production. Of the three types of waste modeled (food waste, unsolicited mail, and drinks packaging), the greatest environmental benefits were seen in food (especially meat) waste prevention.

5.4.2 RECYCLING AND RECOVERY

Food waste may not be suitable for human or animal consumption and its production may be unavoidable. The aim of the recycling and recovery options should therefore be to maximize the value of the various constituents of the food (i.e., nutrients, energy) to make the most efficient use of the resource. Recycling is a more desirable option than recovering as it produces higher quality and usually higher value materials.

5.4.2.1 Recycle

Food waste can be recycled in a treatment process in order to make novel products that can be valuable as soil amendments (i.e., compost and digestate), or it can be used for the extraction of useful chemicals (see Section 5.5.2).

5.4.2.2 Recover

Nutrient or energy value can be obtained from food waste through options such as spreading waste onto agricultural land as a fertilizer or by burning the waste to extract its energy content (see Section 5.5.6).

5.4.2.3 Recycling Olive Mill Waste

Arvanitoyannis and Kassaveti (2007) reviewed the current and potential uses of olive oil waste, a specific problem in the Mediterranean countries of Spain, Italy, Greece, and Portugal where large quantities (estimated at 30 million cubic meters per year) of the waste are produced. The authors found that composting was a promising option for transforming the waste into a useful organic amendment because there were phytotoxicity (due to the presence of polyphenolic compounds and organic acids) and microbial inhibition effects when the waste was spread directly to land. Composting the olive waste, along with other materials necessary to reach adequate conditions for composting, resulted in a reduction in the phytotoxicity but detrimental effects are possible at high application rates (approximately 60 t/ha) due to salinity.

Other than the production of composts, Arvanitoyannis and Kassaveti (2007) suggested that olive waste could be used to produce polymerin—a complex mixture recovered from olive mill waste consisting of carbohydrate, melanin, protein, monosaccharides, phenols, amino acids, and cations. Polymerin may have uses as an agricultural amendment, in biofilters or biointegrators. They also suggested that a three-phase extraction of olive waste could yield sugars and polyphenols that can be used in food preservatives and pharmaceuticals. The waste may also have uses as a substrate for mushroom or microorganism cultivation.

5.4.3   DISPOSAL

Disposal of food is a waste of resources and should be the very last resort for unavoidable waste. In producing a food product, significant resources are utilized—energy, fertilizers, land/soil, water, plant protection products, animal health products, etc.—and wasting the end product by not making the best use of it should be avoided.

5.4.3.1   Incineration

Burning of food waste without utilizing the energy produced or doing so in an inefficient manner is classed as a waste disposal process rather than a recovery process. Incineration produces carbon dioxide and other exhaust emissions (NO_x, dioxins, particulates) that present environmental and human health concerns, and the waste ash from the process must be disposed of.

5.4.3.2   Food Waste Disposal Units

In-sink waste disposal units have been proposed as an appropriate option for food waste disposal, as a means of avoiding landfill of municipal food waste. Although popular in countries such as the United States, disposal units are not widely used in European countries (Iacovidou et al., 2012).

Iacovidou et al. (2012) conducted an economic analysis of food waste disposal units as a means of managing household food waste. Although there were economic advantages to local authorities through the introduction of disposal units in UK households, there were increased costs to water treatment companies in dealing with a new waste stream.

5.4.3.3   Landfill

Landfilling of biodegradable wastes leads to methane and carbon dioxide production for many years after deposition (Donovan et al., 2011). Landfills can also produce leachate that is polluting to groundwater and watercourses, and can cause odor and vermin nuisance problems (Environment Agency, 2009). Landfills are therefore tightly regulated in the EU in accordance with the Landfill Directive (European Union, 1999) and other environmental legislation and should be seen as the last resort for food waste.

5.5   UTILIZING FOOD WASTE AS A RESOURCE

5.5.1   LIFE CYCLE ASSESSMENT (LCA) PRINCIPLES

LCA can be useful for determining the true value and weighing up the options for municipal food waste (Kim and Kim, 2010). When assessing four options for food waste (dry animal feed, wet animal feed, composting, or landfilling) in South Korea, Kim and Kim (2010) used the life cycle expanded system boundary approach to show the impact of each option in terms of CO_2 equivalents. Their approach, based on conditions in one province in Korea, showed animal feed production or composting were better options than landfilling. However, risk impact showed that if the recycled wet feed was not used for its intended purpose (i.e., due to supply and demand issues or quality concerns), incineration of the unused recycled product would be environmentally worse than landfilling. This highlights the need to consider the whole system when assessing the benefits and impacts of food waste diversion policy—the benefits are only felt when the waste is effectively managed.

5.5.2   FOOD WASTE AND BY-PRODUCT UTILIZATION

Traditional forms of waste utilization include use in animal feed or use as a fertilizer (which may or may not involve a treatment stage prior to use). Large-scale food production wastes and by-products offer some advantages over mixed waste streams as they can be utilized more readily in the production of high-value materials. Many of the uses should therefore not be considered waste treatment, as processes can be designed to produce new products or by-products from material that would otherwise have joined a production waste stream. Manufacturing new products or by-products makes economic sense because disposal costs can be avoided by driving a material up the waste hierarchy.

5.5.2.1   Animal Feed

Food can be used for animal feed as an alternative to sending for treatment or disposal. This is normally not considered a waste activity as it actually prevents food waste arising in the first place by diverting it to a different market. There are strict controls on the feeding of animal by-products (or foods that may have come into contact with animal materials) due to biosecurity concerns. In the UK, a strict list of food wastes suitable for feeding to animals includes bakery products (such as bread, cakes, pastry, and biscuits), pasta, chocolate, sweets and similar products such as breakfast cereals; milk and milk products; and eggs and egg products.

The feeding to farm animals of catering waste, kitchen scraps, and raw, partially cooked, and cooked meat products is prohibited under Animal By-Product (ABP) legislation. This legislation is in order to control the potential introduction and spread of viruses and diseases, such as Bovine Spongiform Encephalopathy (BSE), Foot and Mouth Disease (FMD), and Classical Swine Fever (CSF). Outside of the EU, other countries may not have the same rules and sensitivities toward feeding ABP food wastes to animals. For example, in South Korea, food wastes can be recycled into dry or wet animal feed (Kim and Kim, 2010).

Piquer et al. (2009) evaluated whole citrus fruit and citrus pulp as alternatives to wheat grain in the diet of sheep. As well as citrus peel, citrus fruits withdrawn from the market (which may otherwise be wasted) could be suitable for feeding to animals.

Exogenous feed enzymes can be used to enhance the digestion of forage in ruminants and increase production efficiency. Tao et al. (2011) investigated the production of feed enzymes from citrus processing waste by solid-state fermentation (with *Eupenicillium javanicum*). Their results showed that feed enzymes (especially β-glucosidase and xylanase) could be produced by solid-state fermentation of citrus processing wastes and wheat bran.

Bake et al. (2009) evaluated two types of food waste as partial replacements for fishmeal in the diet of Nile tilapia (*Oreochromis niloticus*). They replaced up to 22% of fishmeal with soy sauce waste—the dried and dehydrated residual cake from fermentation of soybeans and extraction of soy sauce—or food industry waste. The food waste included leftovers from convenience stores, food processing residues, hotel waste, restaurant cooking waste, tofu waste, and bread production waste. The food waste was fried in waste vegetable oil and ground to a powder for inclusion in the fish diet. The researchers concluded that these food wastes were acceptable as partial replacements for fishmeal.

5.5.2.2   Biorefinery

The concept of biorefinery of bio-based materials in order to extract high-value products from low-value or waste materials is receiving growing attention. Mahro and Timm (2007) suggested that biowaste from the food industry could be used as a chemical feedstock to produce low volume, high-value products for use in industrial chemicals, cosmetic additives, and gelling agents. They suggested potential products might include collagen, gelatin, chitin, fats and oils, enzymes, flavors and aromas, pigments, antioxidant, pectin, vitamins, and essential oils.

5.5.2.3   Orange Peel

The biorefinery of waste orange peel was recently reviewed by Siles López et al. (2010). Their research estimated world orange production (in 2007) at over 60 million tonnes, with approximately 50 to 60% of this becoming citrus peel waste (consisting of peel, seeds, and membrane) during the process of juice extraction. They discussed the limited existing disposal options for orange peel waste: landfilling or combustion. The potential for utilization of the orange peel waste could be divided into direct utilization as a low-value material, or biorefinery to produce high-value compounds.

Although direct utilization is the simplest method of processing and making use of the raw material, the options are limited to inclusion as a feed for ruminants or application to land as a fertilizer. Siles López et al. (2010) state that the raw material is a good source of nutrients for the microbial communities that exist in rumen, but the waste can develop mycotoxins and degrade if not stored adequately so there is a limit on the proportion that should be used as a feed. Nutrients can also be transferred to agricultural land by direct land spreading or after composting the waste.

Siles López et al. (2010) reviewed the high-value compounds that could be extracted from waste orange peel during a biorefinery process. D-limonene (an essential oil with a strong aroma of orange that is widely used in flavorings and the chemical industry) and pectin (a gelling agent used in the food industry) can be extracted from orange peel. Other products can be generated through the microbial conversion of lignocellulose. These include ethanol (by hydrolysis and fermentation), methane (by AD), pectic enzymes, single cell protein, and vitamins.

Other potential uses for waste orange peel that were suggested by Siles López et al. (2010) included use in the remediation of heavy metal contaminated wastewaters and for the adsorption of phosphate; as activated citrus extract for dermatological treatments and as a preservative; a supplement for paper pulp production; and for the production of the platform chemical succinic acid, which can be used to produce further bio-based chemicals. Mahro and Timm (2007) also suggested that other food industry wastes could be used in the production of platform chemicals.

5.5.2.4   Shellfish Waste

Fishery wastes such as shrimp and crab shells have been exploited for their high chitin contents. Chitin and chitosan are widely used as coagulants, in cosmetics, or in medical materials. To obtain chitin and chitosan from shellfish, first the shell waste must be demineralized and deproteinized using strong acids and strong bases. However, this process also generates waste disposal problems (Wang et al., 2011). Biological methods of chitin production, such as enzymatic treatment and microbial fermentation, are therefore of growing interest. In addition to chitin and chitosan (which is prepared from chitin by deacetylation), Wang et al. (2011) described some new applications whereby bioconversion of chitin-containing seafood processing waste into useful enzymes, biofertilizer, antioxidants, and antimicrobial and antitumor materials is possible.

5.5.2.5   Cheese Whey

Chatzipaschali and Stamatis (2012) reviewed the treatment options for cheese whey—the liquid by-product from cheese-making containing most of the water-soluble constituents from milk. World whey production is estimated at more than 160 million tonnes per year, but only half of that produced is used (Chatzipaschali and Stamatis, 2012; Moulin and Galzy, 1984), although traditionally the whey has been used as animal feed because in the past it was seen as a waste that must be disposed of (Moulin and Galzy, 1984). Options where the whey is treated by a stabilization process or converted into other compounds are attractive because it is expensive to transport liquids.

Whey can be concentrated by reverse osmosis or multiple effect evaporation and spray dried to produce whey powder to facilitate transportation for use as animal feed (Moulin and Galzy, 1984). Proteins can be separated from whey by filtration processes to produce whey protein isolate and lactose (Chatzipaschali and Stamatis, 2012).

These products have applications in food products and pharmacological products. Lactose can also be used to synthesize other derivatives by hydrolysis or fermentation. However, the authors suggest that where the cost of these processes would be prohibitive, AD could be an appropriate treatment technology. AD is seen as a more attractive option because it is a simple, low-cost process with high energy efficiency. In addition to the production of biogas, pollution loadings in the effluent are reduced (Chatzipaschali and Stamatis, 2012).

5.5.2.6 Coffee

Murthy and Madhava Naidu (2012) reviewed the options for recycling the by-products and wastes from coffee production. Different techniques for processing coffee fruit deliver different by-products. The main by-products are coffee husk, peel, and pulp and comprise nearly 45% of the coffee cherry (Murthy and Madhava Naidu, 2012). The authors state that high tannin content prevents coffee pulp from being used extensively in animal feed and the husk and pulp have not been used widely in feed or as fertilizer/compost.

Detoxification of coffee wastes is seen as key to wider utilization, and researchers have used fermentation processes to reduce the toxicity of the material. Murthy and Madhava Naidu (2012) summarized the potential uses for coffee wastes, these included among others:

- fermentation of coffee husk to produce citric and gibberellic acid;
- use of pulp and husk and coffee silver skin as a substrate for production of enzymes and secondary metabolites;
- use of spent coffee and husks for bio-ethanol production;
- production of bioactive compounds, antioxidants, and food additives;
- anaerobic digestion of coffee husk and waste waters to produce biogas;
- use of spent coffee as an adsorbent for waste water treatment;
- use of coffee husks and hulls in the production of particle board.

The authors concluded, however, that commercial processes need to be developed with techno-economic feasibility in order to achieve the potential value-addition and utilization of coffee processing by-products.

5.5.3 Anaerobic Digestion (AD)

AD is a biological process where biodegradable material is broken down by microorganisms in the absence of oxygen. Biodegradable material suitable for AD includes food waste (municipally collected food waste or food industry wastes), animal slurry, or crops grown specifically for the purposes of digestion. There are two products from the AD process: biogas and digestate.

- Biogas: A mixture of 60% methane, 40% carbon dioxide, and traces of other gases. This biogas is combusted to generate heat or power, or enriched to obtain road transport fuel.
- Digestate: A stabilised wet product containing valuable plant nutrients and organic humus. This product can be separated into "liquor" and fiber for application to land or secondary processing.

Treating food waste by AD is a viable option. It is best suited to food waste rather than green garden waste, although green waste can be used to ensure pH stability throughout the process (Akunna et al., 2007). In contrast to compost products, digestate is higher in available plant nutrients and lower in organic matter. If the fibers are not separated from the liquid digestate, whole digestate has been reported at only around 5% solids (Banks et al., 2011). Digestate may require to be separated or dewatered before being utilized; however, it has been shown that dewatering can be reduced by increasing solid content of the digestate by including greater quantities of green waste in the process (Akunna et al., 2007).

AD technology is adaptable and scalable; small-scale simple design digesters can be built for use in developing countries. Viswanath et al. (1992) showed that different fruit and vegetable processing wastes could be fed into a simple 60-L digester to produce biogas and researched the best operating parameters for maximum biogas yield.

AD is becoming an attractive treatment technology for food manufacturers due to increasing disposal costs and the opportunity to reduce costs (Anon, 2012). A lack of facilities for commercial food waste in the UK has meant that many food manufacturers have taken the opportunity to construct on-site AD facilities, as have smaller food businesses such as dairies, distilleries, and breweries (Anon, 2012).

AD has been shown as a viable treatment for the recovery of energy from both source-segregated food waste and the mechanically separated organic fraction from mixed waste (Zhang et al., 2012). These researchers showed that the differing physico-chemical characteristics of mixed and source-segregated waste streams were directly related to the methane potential and the final digestate quality. Both waste types showed good and similar methane yields, but the quality of the mixed waste digestate was higher in potentially toxic elements and lower in nutrients than the source-segregated digestate (Zhang et al., 2012). Chen et al. (2010) investigated five types of food waste as feedstocks for centralized AD. They recommended that food waste should be co-digested with other materials such as manure or meat products in order to minimize the amount of additives that must be added to facilitate the digestion of single-stream wastes.

5.5.4 COMPOSTING

Composting is a biological process where biodegradable material is broken down by microorganisms in predominantly aerobic conditions. Biodegradable material suitable for composting includes green waste (i.e., garden waste), food waste (municipally collected food waste or food industry wastes), and animal manure. The product from the composting process is compost that can be used directly as an agricultural soil conditioner or used in the manufacture of plant growth media.

Although open-air windrow systems are the predominant composting technology within the UK, there has been a shift toward large-scale in-vessel and mechanical biological treatment (MBT) plants in mainland Europe (Sykes et al., 2007).

Food waste can be treated by composting using in-vessel or open-air systems. Food waste (especially that containing animal by-products) is more suited to in-vessel systems rather than open-air windrow techniques. The process of composting food waste (usually in addition to other waste materials such as green garden wastes) produces a soil improver that can be used as a supply of organic matter for soils, as a nutrient source, and to improve soil physical properties.

Unlike AD, the composting process does not produce renewable energy; however, in-vessel systems do have the advantage that they can effectively process food service items such as biodegradable packaging in addition to the food waste (Sung and Ritter, 2008) to produce a soil amendment.

The UK Fertiliser Manual (Defra, 2010) suggests that green waste compost (not containing food wastes) supplies only very small amounts of crop available nitrogen whereas it is indicated that approximately 5% of the total nitrogen applied in green/food compost is available to the crop. The benefits of compost application are therefore through improving soil physical properties through supply of organic matter although soil nutrients may be built up through several years of application.

5.5.5   MUNICIPAL (MIXED) FOOD WASTE

The options for municipal biodegradable wastes are usually limited by the fact that the material is of mixed composition and from multiple sources (households). It is therefore best suited to processes such as AD or composting. AD has a distinct advantage over composting in that it generates biogas in addition to an organic fertilizer by-product.

Russ and Meyer-Pittroff (2004) summarized the potential waste "disposal" routes (excluding the traditional agricultural uses as animal feed and fertilizer) for food wastes: composting, AD, and incineration. There are important considerations for these routes: high water content wastes are unsuitable for incineration but suitable for AD; high cellulose/hemicellulose contents render the waste unsuitable for AD and more suited to composting.

The water content of food waste is generally seen as a problem when exploiting the material as an energy resource (Mahro and Timm, 2007). However, fats and oils can be extracted for utilization, food waste is suitable for the production of biogas by the process of biomethanation (AD), and ethanol can be produced from sugar-containing residues.

5.5.6   ENERGY FROM WASTE

Incinerating food wastes in large-scale plants can be a viable option, although at the lower end of the waste hierarchy and so less desirable as a treatment method. It is preferable to burn the food waste and capture (recover) as much of the energy produced as possible rather than dispose of the material by incineration without energy recovery or landfilling. Energy from waste is most suited to non-source segregated/mixed wastes (which contain variable amounts of food wastes depending on the collection method used locally).

Incineration can generate energy from waste materials, as well as reduce the volume of waste substantially. Incineration or waste-to-energy has been employed widely in Singapore due to land limitations (Khoo et al., 2010). Singapore has therefore adopted the policy of incinerating all "incinerable" solid waste, including food waste; although Khoo et al. (2010) showed using LCA that AD of food waste would significantly reduce greenhouse gas impacts associated with incineration. Composting was also preferable to mass incineration, although it compared less favorably than AD.

By only recovering energy (or a fraction of the energy), this option can be regarded as a waste of the resources used in order to produce food. Japanese law only permits heat recovery from incineration when the food waste is rejected by recycling facilities, due to the poor nutritional value or mixed harmful substances in the food waste or due to lack of local recycling capacity (Takata et al., 2012).

REFERENCES

Akunna, J.C., Abdullahi, Y.A., and Stewart, N.A. 2007. Anaerobic digestion of municipal solid wastes containing variable proportions of waste types. *Water Science & Technology*, 56:143–149.

Anon. 2012. Anaerobic digestion: Turning waste into profit. *Food Engineering & Ingredients*, 37(February/March):15–18.

Arvanitoyannis, I.S. and Kassaveti, A. 2007. Current and potential uses of composted olive oil waste. *International Journal of Food Science & Technology*, 42:281–295.

Australian Government. 2010. *National Waste Report 2010*.

Bake, G.G., Endo, M., Akimoto, A., and Takeuchi, T. 2009. Evaluation of recycled food waste as a partial replacement of fishmeal in diets for the initial feeding of Nile tilapia *Oreochromis niloticus*. *Fisheries Science*, 75:1275–1283.

Banks, C.J., Chesshire, M., Heaven, S., and Arnold, R. 2011. Anaerobic digestion of source-segregated domestic food waste: performance assessment by mass and energy balance. *Bioresource Technology*, 102: 612–620.

Behera, S.K., Park, J.M., Kim, K.H., and Park, H.-S. 2012. Methane production from food waste leachate in laboratory-scale simulated landfill. *Waste Management (New York, N.Y.)*, 30:1502–1508.

Chatzipaschali, A. A. and Stamatis, A.G. 2012. Biotechnological utilization with a focus on anaerobic treatment of cheese whey: current status and prospects. *Energies*, 5: 3492–3525.

Chen, X., Romano, R.T., and Zhang, R. 2010. Anaerobic digestion of food wastes for biogas production. *International Journal of Agriculture and Biological Engineering*, 3: 61–73.

Defra. 2010. *Fertiliser Manual (RB209)*, 8th ed. Norwich: The Stationery Office.

Defra. 2012. Landfill Allowance Trading Scheme Impact Assessment. London.

Donovan, S.M., Jilang Pan, Bateson, T., Gronow, J.R. et al. 2011. Gas emissions from biodegradable waste in United Kingdom landfills. *Waste Management & Research: Journal of the International Solid Wastes and Public Cleansing Association, ISWA*, 29:69–76.

Dorward, L.J. 2012. Where are the best opportunities for reducing greenhouse gas emissions in the food system (including the food chain)? A comment. *Food Policy*, 37: 463–466.

Environment Agency. 2009. *How to Comply with Your Environmental Permit: Additional Guidance for Landfill* (EPR 5.02). Bristol: Environment Agency.

European Union. 1999. Council Directive 1999/31/EC of 26 April 1999 on the landfill of waste.

European Union. 2008. European Parliament and Council Directive 2008/98/EC of 19 November 2008 on waste and repealing certain Directives.

European Union. 2009. European Parliament and Council Regulation (EC) No. 1069/2009 of 21 October 2009 laying down health rules as regards animal by-products and derived products not intended for human consumption and repealing Regulation (EC) No 1774/2002 (Animal by-products).

Eurostat. 2011. *Eurostat Pocketbook: Energy, Transport and Environment Indicators*. Luxembourg: European Commission.

Evans, T.D. 2012. Domestic food waste—the carbon and financial costs of the options. *Proceedings of the Institution of Civil Engineers: Municipal Engineer*, 165(ME1): 3–10.

FAO, WFP & IFAD. 2012. The State of Food Insecurity in the World 2012. Economic growth is necessary but not sufficient to accelerate reduction of hunger and malnutrition. Rome.

Gentil, E.C., Gallo, D., and Christensen, T.H. 2011. Environmental evaluation of municipal waste prevention. *Waste Management (New York, N.Y.)*, 31:2371–2379.

Government Office for Science. 2011. Foresight Project on Global Food and Farming Futures: Workshop Report W4. London.

Gustavsson, J., Cederberg, C., Sonesson, U., Van Otterdijk, R., and Meybeck, A. 2011. *Global Food Losses and Food Waste: Extent, Causes and Prevention*. Rome.

Iacovidou, E., Ohandja, D.-G., and Voulvoulis, N. 2012. Food waste disposal units in UK households: the need for policy intervention. *Science of the Total Environment*, 423:1–7.

Khoo, H.H., Lim, T.Z., and Tan, R.B.H. 2010. Food waste conversion options in Singapore: environmental impacts based on an LCA perspective. *Science of the Total Environment*, 408:1367–1373.

Kim, M.-H. and Kim, J.-W. 2010. Comparison through a LCA evaluation analysis of food waste disposal options from the perspective of global warming and resource recovery. *Science of the Total Environment*, 408:3998–4006.

Kummu, M. et al. 2012. Lost food, wasted resources: Global food supply chain losses and their impacts on freshwater, cropland, and fertiliser use. *Science of the Total Environment*, 438C:477–489.

Mahro, B. and Timm, M. 2007. Potential of biowaste from the food industry as a biomass resource. *Engineering in Life Sciences*, 7:457–468.

Martin, A. and Scott, I. 2003. The effectiveness of the UK landfill tax. *Journal of Environmental Planning and Management*, 46: 673–689.

Moulin, G. and Galzy, P. 1984. Whey, a potential substrate for biotechnology. *Biotechnology & Genetic Engineering Reviews*, 1:347–374.

Murthy, P.S. and Madhava Naidu, M. 2012. Sustainable management of coffee industry by-products and value addition—a review. *Resources, Conservation and Recycling*, 66: 45–58.

Okonko, I., Ogun, A., Shittu, O., and Ogunnusi, T. 2009. Waste utilization as a means of ensuring environmental safety—an overview. *Electronic Journal of Environmental, Agricultural and Food Chemistry*, 8:836–855.

Piquer, O., Ródenas, L., Casado, C., Blas, E. et al. 2009. Whole citrus fruits as an alternative to wheat grain or citrus pulp in sheep diet: effect on the evolution of ruminal parameters. *Small Ruminant Research*, 83:14–21.

Pittman, C. 2010. Portion control: cities attack the problem of food waste. *American Planning Association: Planning*, August/Sep, pp. 22–24.

Quested, T. and Johnson, H. 2009. *Household Food and Drink Waste in the UK*. Banbury: WRAP.

Quested, T. and Parry, A. 2011. *New Estimates for Household Food and Drink Waste in the UK*. Banbury: WRAP.

Russ, W. and Meyer-Pittroff, R. 2004. Utilizing waste products from the food production and processing industries. *Critical Reviews in Food Science and Nutrition*, 44: 57–62.

Siles López, J.A., Li, Q., and Thompson, I.P. 2010. Biorefinery of waste orange peel. *Critical Reviews in Biotechnology*, 30: 63–69.

Sung, M. and Ritter, W.F. 2008. Food waste composting with selected paper products. *Compost Science & Utilization*, 16:36–42.

Sustainable Restaurant Association. 2010. Too Good to Waste: Restaurant Food Waste Survey Report (2010). London.

Sykes, P., Jones, K., and Wildsmith, J.D. 2007. Managing the potential public health risks from bioaerosol liberation at commercial composting sites in the UK: an analysis of the evidence base. *Resources, Conservation and Recycling*, 52: 410–424.

Takata, M. et al. 2012. The effects of recycling loops in food waste management in Japan: based on the environmental and economic evaluation of food recycling. *Science of the Total Environment*, 432:309–317.

Tao, N., Shi, W., Liu, Y., and Huang, S. 2011. Production of feed enzymes from citrus processing waste by solid-state fermentation with Eupenicillium javanicum. *International Journal of Food Science & Technology*, 46:1073–1079.

USEPA. 2011. Municipal Solid Waste Generation, Recycling, and Disposal in the United States: Facts and Figures for 2010. Washington, DC.

USEPA. 2013. Municipal solid waste in the United States 2011 Facts and Figures. Washington DC, available at http://www.epa.gov/epawaste/nonhaz/municipal/msw99.htm

Viswanath, P., Sumithra Devi, S., and Nand, K. 1992. Anaerobic digestion of fruit and vegetable processing wastes for biogas production. *Bioresource Technology*, 40: 43–48.

Von Homeyer, I., Withana, S., Steger, T., Von Raggamby, A. et al. 2011. Final Report for the Assessment of the 6th Environment Action Programme. Berlin.

Wang, S.-L., Liang, T.-W., and Yen, Y.-H. 2011. Bioconversion of chitin-containing wastes for the production of enzymes and bioactive materials. *Carbohydrate Polymers*, 84:732–742.

Williams, P., Leach, B., Christensen, K., and Armstrong, G. 2011. The Composition of Waste Disposed of by the UK Hospitality Industry. Banbury.

WRAP. 2007. Research Summary: Understanding Food Waste. Banbury.

Zhang, Y., Banks, C.J., and Heaven, S. 2012. Anaerobic digestion of two biodegradable municipal waste streams. *Journal of Environmental Management*, 104:166–174.

6 Conclusions

M. Balakrishnan, V.S. Batra,
J.S.J. Hargreaves, and I.D. Pulford

CONTENT

As seen in the preceding chapters, the amount of waste utilized and the extent of development and experience for a given application vary considerably. Of the wastes described, fly ash utilization has a long history for some applications. WEEE, on the other hand, is a relatively new waste. The techniques used to recover metal from electronic waste are similar to primary metal recovery and hence the application is well developed. In the metal processing sector, some wastes like slag from iron and steel production are almost fully utilized, while red mud applications are mainly reported in scientific literature.

With metal processing waste, the high metal content has spurred metal recovery in cases like aluminium dross, where this is a well-established approach. The recycling of salt cake components, however, was motivated by legislation preventing its disposal in landfills. In the case of red mud, probably because of difficulties in its handling due to high alkalinity, large-scale commercial utilization has not been reported although improvements in its storage (e.g., dry stacking) have taken place. The slag from iron and steel, which is generated in large amounts, has a high percentage of utilization while utilization of slag from other metals (copper, nickel) is still in the research stage.

The use of fly ash is primarily in the preparation of cement, concrete, and grout. This constitutes almost 70% of its utilization in some countries like Australia and Japan while in the United States and Europe, it is one-third of its utilization. This application has been recognized since the 1930s. Motivated by the technical advantages and the environmental benefits, its utilization in this application is now well established. Accordingly, standards regarding the use of fly ash in the preparation of these materials have been formulated in various countries including in India and China, which are major producers. The other application contributing to fly ash utilization in the United States and Europe is as structural fill. Another significant waste from combustion is generated during the desulfurization step. The gypsum produced during wet desulfurization has similar properties to that of natural gypsum; hence, it is used as a substitute (e.g., as gypsum panel products). In both fly ash and wet desulfurization wastes, the established applications do not consume the entire generated amount. As a result, research on other applications (such as for waste stabilization) continues.

WEEE recycling is motivated by the recovery of metals including precious metals, which have high value. In addition, availability of some critical metals used for electronic and electrical items is also of concern. The technologies for metal recovery are similar to conventional extraction techniques and recycling facilities are in existence. Informal recycling also exists, especially in developing countries, due to the value of the metals that can be recovered, but this has led to concerns over human health and environmental issues. The generation of large volumes of this waste and the potential toxic components in some parts (lead in CRTs, mercury in LCDs) has prompted legislation in many parts of the world. The legislation covers aspects of collection as well as recovery targets. The challenge is ensuring efficient collection and establishing adequate numbers of suitable recycling facilities, especially in developing economies. Research is also focused on recovery of plastic components and processes for adequate separation of different components in this multicomponent waste.

The reduction and utilization of food wastes is motivated by legislation in many developed countries with restrictions on landfilling of biodegradable wastes. Thus, options such as composting, anaerobic digestion, and use as animal feed are preferred. These approaches are also followed in developing countries due to their economic potential (energy generation in the form of biogas, compost production, etc.). In case of large-scale generation of one type of waste such as in food processing, the option of chemical recovery is being examined. Examples include chitin and chitosan, bioethanol, and platform chemicals. These studies are motivated by the economic value of the recovered products and the savings in disposal costs.

The drivers for waste utilization as described in the introduction are environmental, economic, and legislative aspects. However, the success of any application is linked to good understanding of the application from research studies, the technical and economic advantages it offers, and the extent of legislative enforcement. It is informative to consider the number of publications relating to the various different types of waste to try to gauge the interest in this area within the research community.

6.1 TRENDS IN RESEARCH INTO WASTES

Figures 6.1 to 6.9 show the number of publications over the last 10 years dealing with selected wastes, which have been described in earlier chapters. The data were obtained from Web of Knowledge, using the waste descriptors shown in the figures as the search parameters. Although this provides only a crude measure of the research into the various wastes, it does show some interesting trends and differences between waste types.

Figures 6.1 to 6.3 show the trends for some of the mineral wastes discussed in Chapters 2 and 3. Certain wastes have been utilized for a long time, and so already had significant numbers of publications per year in 2003. A good example of this is fly ash (Figure 6.1), which has long been used for engineering and agricultural purposes. The number of papers on fly ash more than doubled from approximately 800 in 2003 to nearly 1800 in 2012. Steel slag, red mud, and coal combustion

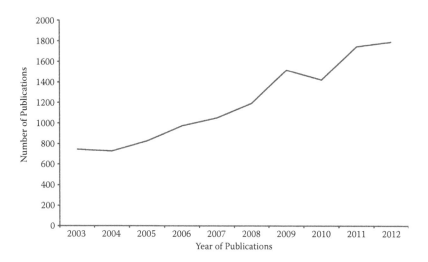

FIGURE 6.1 Annual number of papers published on "fly ash" over the period 2003–2012 inclusively.

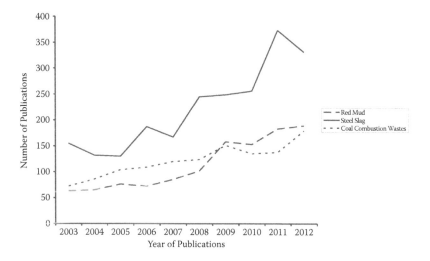

FIGURE 6.2 Annual number of papers published on "steel slag," "red mud," and "coal combustion wastes" over the period 2003–2012 inclusively.

wastes show a similar trend, but at a lower level (Figure 6.2). All of these wastes are produced in large quantities, which may explain the high degree of research interest in them. Wastes produced in smaller amounts—FGD gypsum, FBC ash, and Al dross (Figure 6.3)—all show low-level, fluctuating interest, although the number of papers on FGD gypsum has increased markedly since 2010.

Research papers dealing with WEEE have quadrupled from 23 in 2003 to 92 in 2012 (Figure 6.4), possibly reflecting the increasing legal controls described in

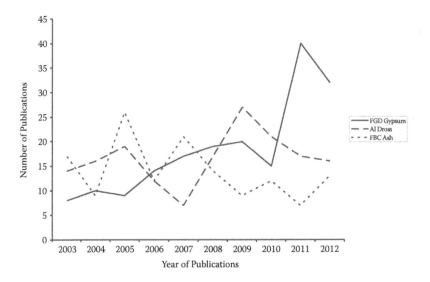

FIGURE 6.3 Annual number of papers published on "FGD gypsum," "Al dross," and "FBC ash" over the period 2003–2012 inclusively.

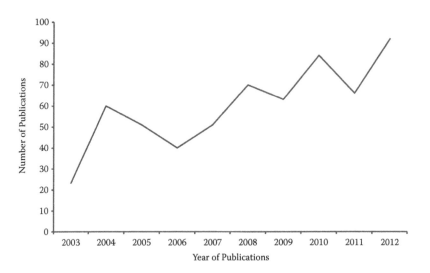

FIGURE 6.4 Annual number of papers published on "WEEE" over the period 2003–2012.

Chapter 4. WEEE represents a slightly different case from the other wastes considered here, in that resource recovery, rather than utilization of the waste, is the more important factor.

As with the mineral wastes, certain organic wastes have been utilized and studied for a long time. Sewage sludge and compost, produced from a variety of starting materials, show trends similar to fly ash, with paper numbers almost doubling from

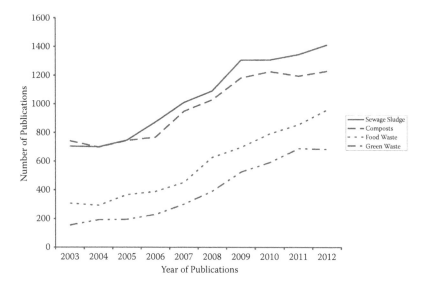

FIGURE 6.5 Annual number of papers published on "sewage sludge," "composts," "food wastes," and "green waste" over the period 2003–2012 inclusively.

approximately 800 to 1400 over the last 10 years (Figure 6.5). Again, both these wastes started from a high base. There has been a much faster increase in the publications on food waste in general, tripling from approximately 300 to over 900, and green waste, which more than quadrupled from approximately 150 to 690 papers. Although some of these papers may also be counted in with composts, the trend does show the specific interest in food wastes and green wastes over the last 10 years.

A comparison of the number of papers on olive mill waste and rice husk ash (Figure 6.6) shows an interesting difference. Both are wastes produced by specific food processing activities, and the number of papers on each doubled between 2003 and 2008. Papers on olive mill waste have since stabilized at approximately 100 per year, whereas those on rice husk ash have again doubled over the last 5 years, to more than 200 in 2012. This may reflect the much greater importance of rice as a global product, and the pressure to find uses for its waste products. Papers on wastes from less important foodstuffs (Figure 6.7) have also increased in number in the last 10 years. The number of those dealing with "shellfish waste" is small (approximately 10 to 20), and has fluctuated rather than showing a steady increase similar to other wastes. Interestingly, if a search is made using "chitosan" as the search parameter there are approximately 5 times as many papers, but the pattern of fluctuation is similar to that for "shellfish waste" (Figure 6.8). Chitosan is of course the main product obtained from shellfish waste.

The category of waste that has shown by far the greatest increase in research interest over the last 10 years is biochar (Figure 6.9). This term covers a range of carbons produced by the pyrolysis of organic materials, and many of the wastes described have been used for this purpose. There has been a steep rise over the last 5 years from almost no publications to over 250 in 2012. This may partly reflect the

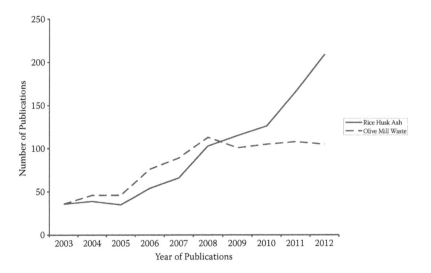

FIGURE 6.6 Annual number of papers published on "rice husk ash" and "olive mill waste" over the period 2003–2012 inclusively.

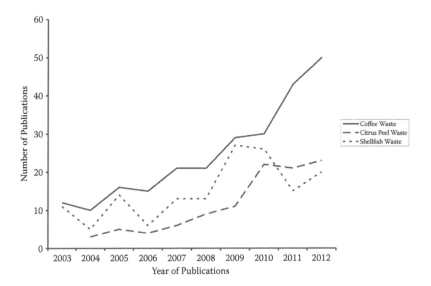

FIGURE 6.7 Annual number of papers published on "coffee waste," "citrus peel waste," and "shellfish waste" over the period 2003–2012 inclusively.

interest in using biochar as a means of carbon capture and storage, as well as utilization of its sorptive properties.

In general, it can be seen that overall the output of research papers relating to waste materials and their utilization is increasing. This reflects the increasing awareness of the scarcity and finite nature of the world's resources and interest in

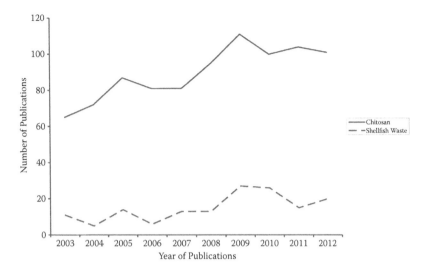

FIGURE 6.8 Annual number of papers published on "chitosan" and "shellfish waste" over the period 2003–2012 inclusively.

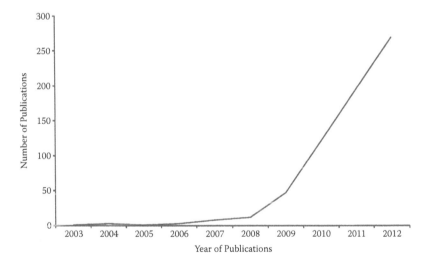

FIGURE 6.9 Annual number of papers published on "biochar" over the period 2003–2012 inclusively.

improving sustainability. Very often large-scale wastes have not only represented a wasted resource but also have presented hazardous situations such as, for example, the storage of high volumes of highly alkaline red mud as exemplified by the tragic events in Hungary in 2010. Wastes and their effective utilization afford interesting, and important, opportunities and interesting future developments in this vibrant and increasingly recognized area can be anticipated.

Index

Milton Keynes UK
Ingram Content Group UK Ltd.
UKHW031151141024
449569UK00024B/887